SACRED GEOMETRY

A
CONSCIOUS
GUIDE

JEMMA FOSTER

SACRED GEOMETRY

How to use cosmic patterns to power up your life

aster

To Maya

An Hachette UK Company
www.hachette.co.uk

First published in Great Britain in 2020 by Aster,
an imprint of Octopus Publishing Group Ltd, Carmelite House,
50 Victoria Embankment, London EC4Y 0DZ
www.octopusbooks.co.uk

Distributed in the US by Hachette Book Group
1290 Avenue of the Americas, 4th and 5th Floors, New York, NY 10104

Distributed in Canada by Canadian Manda Group
664 Annette St., Toronto, Ontario, Canada M6S 2C8

ISBN 978-1-78325-341-8

A CIP catalogue record for this book is available from the British Library.

Printed and bound in China.

10 8 6 4 2 1 3 5 5 7 9

Consultant Publisher: Kate Adams
Art Director: Juliette Norsworthy
Senior Editor: Leanne Bryan
Production Manager: Lisa Pinnell
Copy Editor: Marion Paull
Designer: Rosamund Saunders
Illustration: Joel Galvin
Graphic Design: Andreas Brooks & Maximillian Perchik
Front & Back Cover Illustrations; Chapter Mandalas: Jemma Foster

The information given in this book is not intended to act as a substitute for medical treatment.

Contents

Introduction

'Philosophy is written in that great book which ever lies before our
eyes – I mean the universe – but we cannot understand it if we do not
first learn the language and grasp the symbols, in which it is written.
This book is written in the mathematical language, and the symbols
are triangles, circles and other geometrical figures, without whose
help it is impossible to comprehend a single word of it; without which
one wanders in vain through a dark labyrinth.'

Galileo Galilei (1564–1642), astronomer, physicist and engineer

Sacred geometry is the language of the universe. It is
the creative formula that breathes life into existence.
Its energetic architecture reveals itself to us through form,
pattern and number. It provides the vibrational blueprint
for the construction of matter, and we can witness these
archetypal designs at work in everything from the subatomic
to the galactic. It is reflected in the division of our cells,
the unfurling of a rosebud and the orbit of the planets.
Geometry is about relationships. It describes the way in
which any form – animate or inanimate – measures up to
itself or another. It is not an invention; it is inherent in
all things. Rooted in its nature is the understanding that
nothing is in isolation, everything is connected.

You do not have to go anywhere to find it. It is your mother tongue, the language of your atoms and your cells. It is the intelligent order that speaks through the design of nature. It is there in the hexagonal pattern of a honeycomb, the five-pointed star of an apple core and the spiral of a pine cone. It is the way your embryo was formed as you were in your mother's womb, the relationship of your belly button to your big toe, the architecture of the building that you occupy. It is the distance from you to the Moon and the distance of the Moon's orbit around the Earth.

DERIVATION OF THE WORDS

Geometry comes from the Greek *geometria* meaning 'measurement of the earth or land', *gē* – earth, *metria* – measure. In Olde English it is *eorðcræft*, meaning 'earth-craft'.

To be 'sacred' is to be worthy of reverence. It stems from the Latin *sacrare*, from *sacer*, *sacr-* meaning 'to sanctify', or that which is 'holy', and to be holy is to be whole. Holy is from the Olde English *hál*, meaning 'in one piece', 'unharmed', 'complete'.

The physical reality that we see is the result of a complex set of interactions, correspondences and flirtations that begin on unseen levels beyond our perception. At the dawn of our awareness, there is a conscious, organizing principle that defies understanding by science and logic alone. It is not a consciousness that is a mere product of the brain but one that is primal and expansive, a generative life force at work weaving together the fabric of the universe.

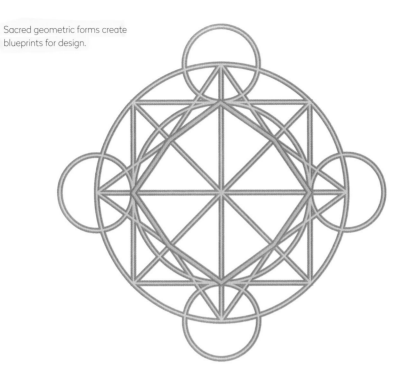

Sacred geometric forms create blueprints for design.

'What is it that breathes fire into the equations and makes a universe for them to describe?'

Stephen Hawking (1942–2018), theoretical physicist

To understand this igniting source is the eternal quest at the heart of humanity. The desire and obsession to unlock its secrets has been encrypted in our art, architecture, religious symbolism and scientific endeavours throughout the ages.

Ancient calendars were based on natural growth patterns and cosmic cycles. The Mayan civilization devised a complex astrological system, of which the Tzolkin (below) is just one of three calendars used for divination.

Pattern recognition

Sacred geometry is the union of pattern and symbolism.
Our survival on this planet has been dependent upon pattern
recognition. Our ancient ancestors observed nature and
watched the sky so that they could predict the seasons and
eclipses, navigate the tides for fishing, know when to gather
wild plants and when to sow and harvest crops. They clearly
perceived the connection of the planets with their own psyche
and biological cycles, such as the female menstrual cycle
or Moon Time, and created ceremony in response. This
understanding led to the practice of astronomy, astrology
and the building of temples and monoliths to track and
mark these crucial points throughout the year.

Wisdom is communicated through symbolism. Psychoanalyst
Carl Jung (1875–1961), in his book *Man and His Symbols*, said
that a word or image is transformed into the symbolic when
it has a hidden, unconscious meaning that requires decoding.
Symbols are tools of expression and keys to knowledge.
Their use is how we learned to interpret the world around
us and to pass on knowledge to future generations through
the development of language, music and symbolic art.

Sacred geometry is your story too, how you came into being
and how you continue to evolve. You already have a deep,
primal relationship with sacred geometry whether you realize
it or not. It is in the world around you, everything that you
interact with and how you express yourself through sound,
movement and energetic exchange. Your soul recognizes it
and the process of discovery you are about to embark on is
one of remembering.

How to use this book

This guide is intended to take you on a journey into the ancient mystical teachings of the past. It is an invitation to return to nature, the cosmos and our ancestral ways by looking at the building blocks of existence through the language of sacred geometry.

Sacred geometry is a manifestation tool. It bridges the gap between our earthbound reality and other dimensions of perception. These vibrational shapes shift our frequency field to allow for the right conditions to bring about a change in our direct experience. When we are existing in accordance with these sacred principles, we are in a state of abundance and receptivity. We become the co-creators of our reality.

FEED THE INTELLECT

From the mystery schools of Ancient Egypt and Greece, where initiates would be taught the secrets of the universal order, sacred geometry developed as an academic art that combined the traditional philosophical practices of number, geometry, music and cosmology, known as the quadrivium. The intellectual mind is a wilful force and often needs satisfying with such study before it will release its grip, so if you have a strong mind, it will relish the academic nature of the subject, but it is to the heart that this language truly speaks. It is non-linear, outside of space-time and the limitations of the physical spectrum.

I have painted a series of mandalas as introductions to the chapters. Each one carries the energetics of the information contained within its chapter and serves as a supportive device. Take a moment to tune in before reading on.

As we move through the chapters, there will be practical applications including meditations, visualizations, divination and crystal grids.

In the Meditative Geometry chapter (*see* page 143), there are additional step-by-step instructional drawings on how to draw the shapes for yourself. Something happens on a very profound level when we engage in creating these images by hand. Each line or curve draws us closer to home, to our inner still point. The process of drawing these shapes is intuitive, sensual and deeply meditative. It is a transcendental experience, connecting us to the sacred proportions of the universe.

USE THE SHAPES

If you would like to engage with the shapes in this book and use them in your daily life, you can photocopy or trace any of the images featured and colour them in as a meditation. Use them as templates for crystal grids, or place them in your home to anchor their energies into your space.

Movement and form

'Everyone who is seriously involved in the pursuit of science
becomes convinced that a spirit is manifest in the laws of
the universe – a spirit vastly superior to that of man, and
one in the face of which we, with our modest powers, must
feel humble.'

Albert Einstein (1879–1955), theoretical physicist

Sacred geometry is an energetic language that
communicates consciousness and influences matter
through fluctuations in vibration. It is worth spending
some time exploring what this means, looking at how
the universe works and the role of consciousness
within matter.

Consciousness

'If you want to find the secrets of the universe, think in terms of
energy, frequency and vibration.'

Nikola Tesla (1856–1943), engineer and inventor

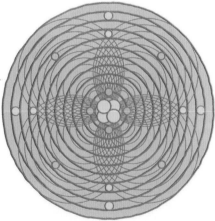

Atomic vibrations, matter
and the quantum field.

The universe that we are concerned with
is not the mechanical, clockwork version
of Sir Isaac Newton but the living,
interconnected field or energy matrix
that underpins esoteric philosophy and
quantum physics. As for consciousness,
it is the creative spirit at work, the
mysterious glue that holds our universe
together. It manifests living form, capable
of extraordinary and unfathomable acts.

It has been called God, the creator, the void, the cosmic sea, dark matter, gravity, the unified field and a host of other things. It is what Buddhist and Vedic texts refer to as a void pregnant with creative potential, from which everything arises and passes. It is the nothingness and *everythingness* of existence. It is the dance between realities, between the seen and unseen.

Einstein's first law of thermodynamics states that energy cannot be created or destroyed, it can only be changed from one form to another. This can be a constructive role – like the anabolic creation of new cells in our body – or destructive, mirrored in the catabolic breakdown of waste products in our body. In Eastern philosophy energy is called *qi* or *prana* and refers not just to the production of physical change but to the life force or spirit of the conscious universe itself. Boundless energy exits in the universe, the magic lies in the ability to harness and redistribute it.

Vibration and frequency

Everything in the universe vibrates. It is in a constant state of flux and motion. Even the molecules in the chair you may be sitting on are vibrating at a colossal rate. Whether we're talking about a plant or the pot it sits in, both have their own vibrational frequency and it is this that creates the appearance of matter. The way in which something vibrates dictates its density, form and how it behaves around gravity.

The speed at which these subatomic particles are jostling around can be measured as frequency, that is the number of wavelengths (the distance between the crest of one wave and the crest of the next) passing through a window in a set time frame (measured in metres per second). The entire universe is made up of waveforms and it is by observing the relationships between them and giving them numerical value that we begin to 'know' our galactic home. The higher the frequency, the closer the waves are together and the more energy generated. Lower frequencies create longer wavelengths.

Oscillating wave forms are measured in frequency.

ELECTROMAGNETISM
Light consists of oscillating fields of electricity and magnetism. It is all around us and within us, stretching far beyond our visible horizon of perception; even our DNA beams out biophotons (tiny packets of light).

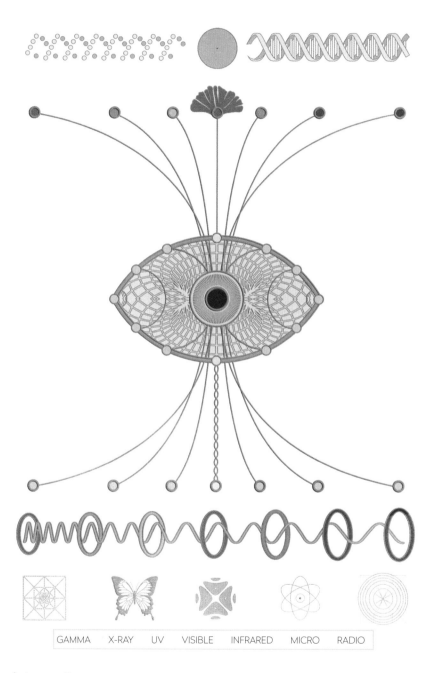

GAMMA   X-RAY   UV   VISIBLE   INFRARED   MICRO   RADIO

The electromagnetic
light spectrum.

Matter

'The physical body is made up of intersecting, pulsating energy fields. What we call a 'physical body' – flesh, bones and blood – rapidly disappears when highly magnified. A physical body, therefore, or any piece of matter, can be viewed as an interference pattern of electromagnetic fields that change with the passage of time.'

Itzhak Bentov (1923–1979), scientist, inventor and mystic

Each vibrational frequency creates the blueprint or sacred geometric design to form an energetic structure. A wave, when crossed with another wave, creates an interference pattern which gives us matter. This is why physicist David Bohm (1917–1992) called matter frozen light. It is because waves of energy move a certain way that we have particles and a solid object as a result, but if the vibration changed and sped up phenomenally, the object would no longer be solid.

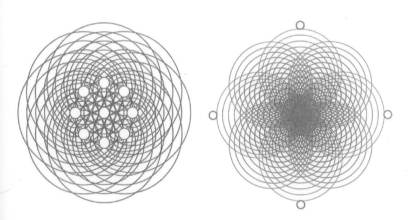

Interference patterns: matter as frozen light.

Water illustrates this perfectly. When water is frozen, the molecules inside it are vibrating very slowly making the water dense and solid. As it is heated up, and the particles begin to vibrate faster, they move apart, expanding to liquid and then eventually steam. In reverse, steam condenses into liquid and solidifies into ice. The principle is the same with sacred geometry –the vibrational blueprint creates the conditions in a non-physical reality and condenses it into the physical, material world as a solid. Flowing water has been found to alter its geometry in this way to maximize flow and navigate the landscape and the challenges of gravity. This is the dance of consciousness and creation, from spirit into matter and back again.

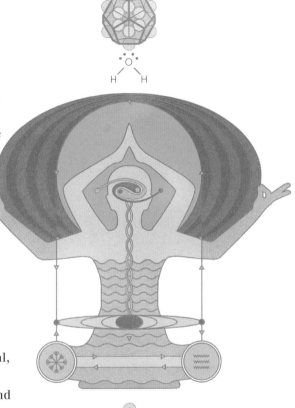

An esoteric rendering of the geometric structure, balance and flow of water and its three states – solid, liquid and gas – superimposed on the human form.

Atoms

The Greek *atomos,* meaning 'not cut' or 'indivisible', is the root of the word 'atom' because atoms were once believed to be the smallest components of our universe. Atoms are formed of a nucleus containing subatomic particles called neutrons and protons, with orbiting electrons. If you think of an atom as an orchestra, then neutrons and protons would be the instruments and the electrons would be the conductors.

Quantum physics has upgraded our take on this to understand that particles can also behave as waves and that things get much, much smaller than we thought (the smallest known elementary particle is the quark). What's more, when you dive into an atom it is actually 99.9999999999999 per cent 'empty space.' If you removed this 'empty space' from the atoms of all the people alive on Earth, the entire human race could fit inside a sugar cube. However, the space left over is far from empty – it is known as 'vacuum energy', which is the energy that exists in the 'background' throughout the universe, a mysterious ocean of potential from which everything arises and passes into existence.

Atomic structure and vibration.

It is from this space that electromagnetic waves crash onto the shores of our three-dimensional reality as particles contributing to matter. Physicists Richard Feynman (1918–1988) and John Wheeler (1911–2008) calculated that a single light bulb's worth of this energy would be enough to boil all the world's oceans.

An atom, then, can be thought of as a complex web of waveforms, an energetic matrix in a constant state of flux and transformation potential. If geometry can act as a sort of interdimensional glue holding stuff together, then ⌈consciousness is this unified field of potential.⌋ It is the 'empty space' in our atoms and the mysterious substance that grips our universe together. Noble Prize-winning father of quantum mechanics Max Planck concluded: 'There is no matter as such. All matter originates and exists only by virtue of a force which brings the particle of an atom to vibration and holds this most minute solar system of the atom together. We must assume behind this force the existence of a conscious and intelligent mind.'

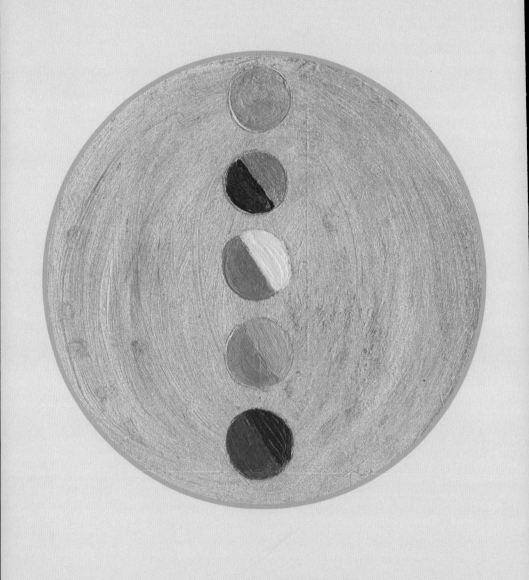

Creation

'The nitrogen in our DNA, the calcium in our teeth, the iron in our blood, the carbon in our apple pies were made in the interiors of collapsing stars. We are made of star stuff.'

Carl Sagan (1934–1996), astrophysicist

As the rhythm of creation, sacred geometry gives us a symbolic reference for the birth of the universe. Creation myths usually begin in silence and darkness. All potential existence is held in a cosmic void or primordial sea from which some great being of creation stirs out of a deep sleep, expands its awareness, and awakens the universe with a sound, the sacred om. Vibrations ripple through the cosmos and light fractures across time and space, structuring our universe with intelligent design through geometric form.

IMAGINATIVE BEGINNINGS

The cosmic egg cracks open, a serpent uncoils its body, a lotus flower blossoms. Often there is an implied prior existence, the sculpting of the universe as part of a wider cyclical evolution and creator beings bridging or transitioning from one world to the next. It is from the imagination of this being or creature that our universe is conjured up and inhales its first, deep, cosmic breath.

THE STORY OF CREATION IN 2D

●   Dot: zero dimension, the singularity

—   Line: movement between one point and another

𝓁𝓁   Spiral: spin, vortex energy, motion, electromagnetism

○   Circle: wholeness, oneness, completion and beginnings

◎   Double circle: duality, opposing and equal forces, attraction

△   Triangle: connection, integration, balance

□   Square: manifestation, matter, grounding, stability

Dots, lines, waves and spirals

A dot is often referred to as dimensionless – it has no beginning or end but is at once the end and the beginning. It is mysterious and unknowable. Like a seed, it holds all future potential within it – it is the part containing the whole; a grain of sand that forms the Sahara; a bindi painted on a forehead carrying a message to the gods; a pupil contracted in sunlight. It heralds the journey of a life and it concludes the end of this sentence.

A line is a point moving through time, from one place to another. You could think of it as a continual series of dots marching forwards or backwards. Sound or vibration is measured as a wave or moving line. As the conscious entity moves, it oscillates outwards forming a spiral. We see this process in our own birth. The sperm tail is an oscillating wave that penetrates the egg. Once fertilized, the egg splits in two and the process of creation begins. Embryology unfolds in a spiral formation. Mitosis is the name for this cell division, and *mitos*, like line, *linea*, means 'thread'.

STRING THEORY

Think of waves and particles as lines and dots. This is essentially what we are working with using sacred geometry. Everything in the universe has some degree of curvature. Lines are never really straight lines; they are waves, and circles are spheres. In Hindu texts, a circle is described as an illusion because in reality, all circles spin in spirals. A major idea in quantum thought is string theory, which suggests that the web of the universe is woven from strings of energy. This is a concept that goes back a long way and is reflected in the beliefs of the Ancient Egyptians (whose goddess Neith wove together the tapestry of the universe) and the Hopi and Navajo in North America (who look to Spider Grandmother, who casts a web of protection and wisdom).

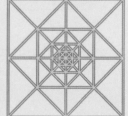

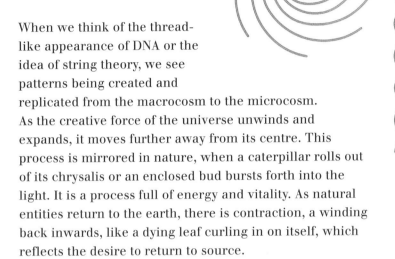

Galaxy spiral (right) and DNA (far right).

When we think of the thread-like appearance of DNA or the idea of string theory, we see patterns being created and replicated from the macrocosm to the microcosm. As the creative force of the universe unwinds and expands, it moves further away from its centre. This process is mirrored in nature, when a caterpillar rolls out of its chrysalis or an enclosed bud bursts forth into the light. It is a process full of energy and vitality. As natural entities return to the earth, there is contraction, a winding back inwards, like a dying leaf curling in on itself, which reflects the desire to return to source.

VISUALIZATION: CREATIVE IMPULSE

Try this: visualize a pool of mirror-like water. This is consciousness, the primal source, the reflection of our mind when undisturbed by the vibrational frequency of our thoughts and emotional patterns. The impulse to create is like dropping a pebble into that water. The vibrations ripple outwards from the centre until they reach the boundary conditions for creation – the unique design template that distinguishes one thing from another, that makes a frog a frog and a rock a rock. This demonstrates the process of oscillation and radiation in the creative principle. As the waves move from the centre, they become denser. Earth has a gaseous liquid core and hard crust, just as we have gaseous liquid inside of us wrapped in a layer of tight, solid flesh, and marrow held inside our bones. Plants have hollow stems and fruits have air pockets full of seeds inside.

Ouroboros/
Chrysopoeia
(top far left); Reiki
Cho-Ku-Rei
symbol (bottom
far left). Orphic
Egg (left).

Circles and spheres

In many cultures, the origin of life is represented as a circle, sphere, cosmic egg or a serpent eating its own tail (the ouroboros or chrysopoeia). It is representative of pure consciousness, timeless and eternal. It is a symbol of perfection, wholeness, oneness and completion. It represents the cyclical and rhythmic pattern of life, death and rebirth.

Yet perfect circles rarely appear in nature – planets take wobbly and erratic orbits; cell membranes are forever-morphing bubbles, expanding and contracting; up close, the Sun is a jagged ball of flames and the Moon is an undulating landscape pitted with craters. There is a yearning to achieve that sacred sphere and everything is moving towards it, desiring to imitate its perfection. In sacred geometry, we are concerned with perfect shapes because they are representative of a pure, idealized form. This archetypal language bridges the gap between the perfect unity of spirit

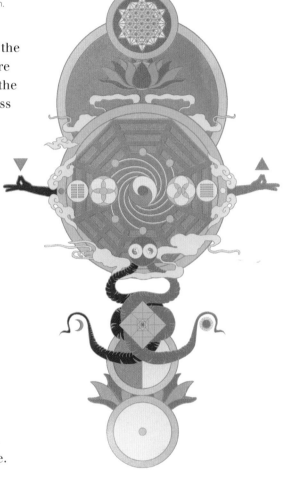

An expression of Tao cosmology and the Eight Trigram Furnace where the Pill of Immortality was fired. The duality of yin and yang are also the short and long lines that create these *Bagua* – trigrams relating to the principles of creation.

and the duality-consciousness of the physical realm. These patterns are divine, uncorrupted, illustrating the pivotal moment when the formless enters into form.

From this place of unity and wholeness, the circle or sphere must divide, split itself in two in order to create another. This is what we call duality, it is a move away from wholeness into otherness. In it we have equal but opposing forces of yin and yang, light and dark, masculine and feminine – all parts of the whole. At the same time, the circle is the seat of creation. From the interference pattern of two oscillating wave forms – like the ripple of water disturbed in a pond – emerges a third, new state. The parents give birth to a child.

'Tao gives birth to One, who produced Two and Two produced Three. These eventually created all other beings.'

Lao Tzu (born *c.*601 BC), philosopher and writer

VESICA PISCIS

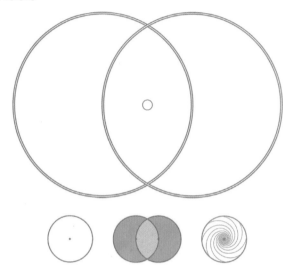

The two overlapping circles of the Vesica Piscis create the lens of the Mandorla (left); the process creates an energetic vortex of creation (bottom).

The Vesica Piscis is formed from this moment of creation. It is the basis of a series of fundamental sacred geometric shapes and patterns that present the creative force behind life. It is the exchange of information between two entities to create a third paradigm. At the point of intersection, a lens forms in the shape of an almond – called the Mandorla.

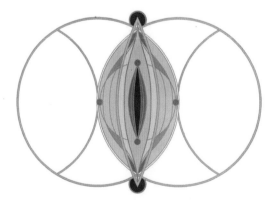

Vesica Piscis means 'vessel of a fish' and is called this because it resembles the conjoined double bladders of many fish. It is also considered to represent the vulva, or *yoni*, of the Goddess, Mother Earth – the female organ of creation through which the universe is born.

SACRED PORTAL

Yoni in Sanskrit means 'holy passage', representing the door for the soul to enter the physical world. It is the sacred space of harmony and creation and offers a lens of perception through which to see the world. The Vesica Piscis can be found on the door to the Chalice Well in Glastonbury and in the relationship between the proportions of the Sphinx and the Great Pyramid of Giza. It is considered a portal between worlds, the bridge between the physical and non-physical realms.

A solar eclipse is a natural display of a Vesica Piscis. Through this we can see the dance of duality at play, as the opposite forces of dark and light converge. In many ways, an eclipse is a call to embrace our own shadow side and carve a new path from the acceptance of it. In Freudian terms, the shadow is the darkness that lurks within each of us, whereas Jung saw it less as a negative construct but more the sum total of what is hidden from our awareness. Shadow work is an invitation to hold a mirror of truth to ourselves and meet our reflection with compassion and forgiveness for the fallibility that is inherently human.

VISUALIZATION: FACE YOUR SHADOWS

See where there is duality within yourself, feel into any imbalances or opposition that you are holding on to and see where there might be resolution. Invite your shadow self into the light without judgement or intention, just acceptance that it is part of the whole of you.

REBALANCING YOURSELF MEDITATION

The Vesica Piscis is a tool for addressing and balancing any duality in your life and for calling in the forces of creation. The circle on the left can be representative of the feminine, the past and the emotional body supported by the Moon. The circle on the right is the masculine, the future and the physical body supported by the Sun. The Mandorla is the point of unity, the merging of these opposing forces, the sacred space of harmony and creation.

This page and opposite: Vesica Piscis, creation and the sacred spiral.

- Set yourself up for your meditation. You can do this anywhere at any time.
- Stabilize your nervous system by taking six slow, deep breaths in through your nose and out through your mouth.
- Root from your feet into the earth and draw its vital force up to your hips and into your heart.
- Visualize the two equal circles of the Vesica Piscis and what they represent to you in this moment.
- Roll them outwards simultaneously – the left to the left and to the past and the right to the right and to the future. As you do so, imagine that each circle is returning what you are holding on to in the present to where it belongs in the past or the future.
- Once you have reached the limit of your timeline, as far as you are permitted in this session, roll the circles back to the centre, collapsing all alternate realities and coming fully into the present moment.
- Where the circles intersect to form the Mandorla is the present moment, the now, and it represents you. What do you want to manifest into your reality now? What will you create through unity?

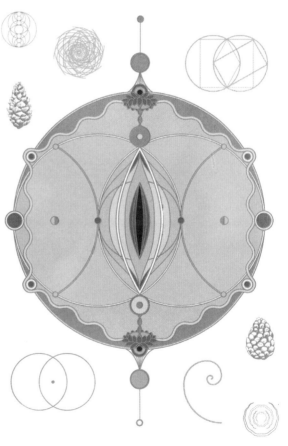

LINE DIRECTION

A vertical line can be seen to represent the transformation of energy from spirit descending into matter and the rising of human consciousness towards spirit. A horizontal line signifies the unification of the soul energy between two entities, dimensions or realms.

CIRCLES AND LINES

Circles are considered feminine and straight lines masculine, and it is by harmonizing these energies that we can reveal the secrets hidden within. This is the true nature of sacred geometry. A wealth of information is encoded within the Vesica Piscis, which can be unlocked by adding straight lines (*see* illustrations, opposite).

To create a Vesica Piscis, first draw a circle using a compass (*see* images, top right and centre right). Maintaining the width of the compass, draw a second circle from the circumference of the first (*see* image, bottom right). This duality state is the fertile ground for creation and from it many other shapes can be born.

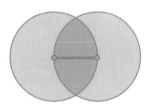

How to draw a Vesica Piscis, step by step.

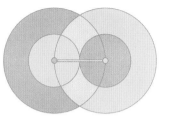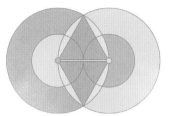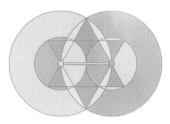

Adding straight lines to the
Vesica Piscis can unlock
hidden meaning.

JOINING THE DOTS

We have a dot and a line moving away from it that comes
to an end with another dot. The two dots are the distance
of the line apart from each other. To unify them, we need to
create a third dot and join them together to form a triangle.

This unity is reflected in the holy trinity of many cultures
in which there is a formless higher creative consciousness,
expressed through a spirit that mediates with and expresses
itself through the image of humankind. Within this
triangular relationship there is the one, the one and the two,
and the one, two and three. One is the consciousness that
contains all possibilities within it. Two is humankind; it is
equal and opposite forces, duality and unity, attraction and
repulsion. Three is the joining together of one and two,
through a third source or mediator.

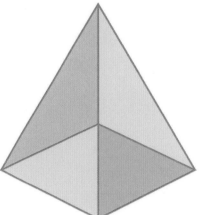

A pyramid, consisting of a square base with four triangular sides.

Triangles and squares

From the triangle, we can create an equilateral tetrahedron with four equal triangular sides. With the addition of the square, we can make a pyramid (as above) consisting of a square base with four triangular sides. The number four, represented by the square base, signifies all that is manifest in our physical reality. It is the end point of the initial stage of creation, when matter is formed and a stable earthbound state is reached. It represents the world as we know it. The idea of three inside four, or the triangle inside the square, is repeated throughout alchemical and sacred geometric lore as representative of spirit (non-physical) inside man (physical).

Another shared concept in creation myths is that the process of building the universe took six days plus a seventh day of rest or completion. The Vesica Piscis represents the first two days of creation, after which the Seed or Egg of Life is formed, containing seven circles. From this develops the Flower and Fruit of Life. With each progression more knowledge about the complexities of existence are woven into the design. These primal patterns are found throughout the world, always bearing the same names.

Clockwise from top: Flower of Life, Seed of Life, Fruit of Life, Egg of Life.

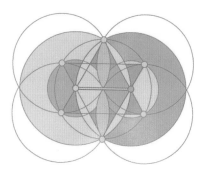

SEED OF LIFE

▲ Take your compass and place the point on the page. Think of the compass point as your dot of creation. This is your still point, your centre. Stretch out its arms and arch its back – this is your spine, the axis on which you pivot. Like a dancer, leg outstretched, ready to pirouette. This represents the first light, first word, the first movement out of the stillness.

▲ Draw your circle, creating a boundary. Observe yourself in this process. Consider the state of your inner compass, the refinement of your centre and the radius of your energy field. What level of consciousness are you sending out into the world and how fine-tuned is your antenna?

▲ Maintaining the width of your compass, place its point at the edge of the circumference of the circle and draw a second circle to the right of the first. This is your Vesica Piscis, days one and two of creation.

▲ Now place the compass point at the lower intersection of the two circles and draw a third circle to create a trinity.

▲ Repeat the process moving clockwise to draw four more circles at the intersections of your existing three, so that you have seven circles and a six-petal flower appears in the centre.

This is the Seed of Life, the blueprint for creation and for that reason, it is called the Genesis pattern. The whole is contained within the seed – the entire future potential of a plant is contained within its geometry like a hologram, and all the plants that came before it are programmed into the seed. We see the same holographic nature – every part contains the whole – in a fertilized egg.

The evolution of the
Seed of Life.

EGG OF LIFE
The Egg of Life is a three-dimensional
form of the Seed of Life which mirrors
an embryo in the first hours of creation.

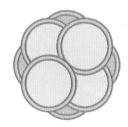

The progression from
seed (left) to egg (right).

THE FLOWER OF LIFE

The Flower of Life is the cosmic generator. It supports and sustains all life. In mystic texts, it is said to be an energy matrix of information that channels pure consciousness and holds the Akashic records – the etheric library containing the records of all existence. Information, like energy, cannot be destroyed and is imprinted into the collective unconscious. The Flower of Life is found on the 5,000-year-old walls of the Osirian temple in Abydos, Egypt, where it is burnt into the stone as if by laser. It is also under the paw of the Fu-dog guarding the Forbidden City in Beijing and in countless other temples and sacred sites across the globe.

To create the Flower of Life we expand on our Seed of Life by first adding another six circles at the intersections of the original circles. Then add a further six outer circles to those intersections. Then add arcs to complete the remaining petals of the flowers within the outer circle. This is then enclosed by a circular membrane.

TREE OF LIFE

The Tree of Life is contained within the Flower of Life and represents the whole cosmology of creation. It is a spiritual map of the universe and illustrates the relationship between consciousness and matter.

The Tree of Life superimposed onto the human form (right) showing the ten spheres of the sephiroth and the junction of Da'at, meaning knowledge, at the head; the Tree of Life in the Flower of Life (above right).

FRUIT OF LIFE

We find the most precious Fruit of Life hidden within the Flower of Life, revealing thirteen circles signifying the abundance of creation and transition between dimensions. It is the metaphorical and literal fruition of our journey.

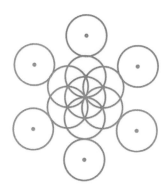

METATRON'S CUBE

If we add straight lines to connect the circles of the Fruit of Life, we create Metatron's Cube. Metatron is described in mystical texts, such as the Babylonian Talmud and in the Jewish Kabbalah, as Archangel Metatron, the guardian of human life force and the overseer of the Tree of Life. He is the gatekeeper of the Keter or crown chakra through which consciousness flows. In Egyptian mythology, he is associated with the god Thoth, the messenger and scribe of life.

Elements contained within Metatron's Cube (top)
and the Platonic Solids (bottom; *see* page 46).

The Platonic Solids

Contained within Metatron's Cube, are the Platonic Solids, named after the Ancient Greek philosopher Plato (*c.*427–347 BC), who saw them as the fundamental building blocks of life. They are regular convex polyhedra (many sided) – three-dimensional shapes with polygonal (two-dimensional) faces. Each solid represents an element – earth, air, fire, water and aether. Aether – also 'ether' – is the spirit, the life force and consciousness that pulls the other elements together and connects the physical and non-physical. In 1986 physicist, chemist and engineer Dr Robert J Moon (1911–1989) confirmed that the Platonic Solids correlate to the atomic structure of the entire periodic table.

The Platonic Solids. Clockwise from left: fire, earth, air, aether and water.

THE PLATONIC SOLIDS

tetrahedron	4 sides	4 equilateral triangles	fire	creation, inspiration, sexuality
hexahedron (cube)	6 sides	6 squares	earth	grounding, receptive, sensual
octahedron	8 sides	8 equilateral triangles	air	communication, breath, nourishment
dodecahedron	12 sides	12 equilateral pentagons	aether	life force, spirit, cosmic consciousness
icosahedron	20 sides	20 equilateral triangles	water	flow, emotions, boundaries

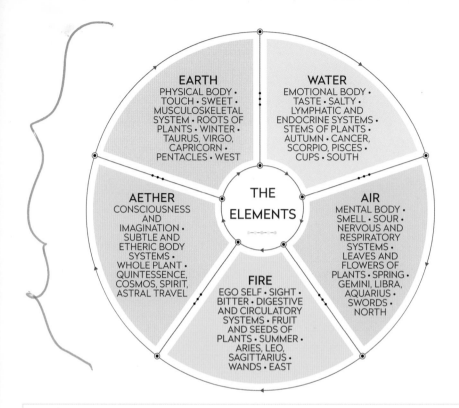

THE ELEMENTS

EARTH
PHYSICAL BODY ·
TOUCH · SWEET ·
MUSCULOSKELETAL
SYSTEM · ROOTS OF
PLANTS · WINTER ·
TAURUS, VIRGO,
CAPRICORN ·
PENTACLES · WEST

WATER
EMOTIONAL BODY ·
TASTE · SALTY ·
LYMPHATIC AND
ENDOCRINE SYSTEMS ·
STEMS OF PLANTS ·
AUTUMN · CANCER,
SCORPIO, PISCES ·
CUPS · SOUTH

AETHER
CONSCIOUSNESS
AND
IMAGINATION ·
SUBTLE AND
ETHERIC BODY
SYSTEMS ·
WHOLE PLANT ·
QUINTESSENCE,
COSMOS, SPIRIT,
ASTRAL TRAVEL

AIR
MENTAL BODY ·
SMELL · SOUR ·
NERVOUS AND
RESPIRATORY
SYSTEMS ·
LEAVES AND
FLOWERS OF
PLANTS · SPRING ·
GEMINI, LIBRA,
AQUARIUS ·
SWORDS ·
NORTH

FIRE
EGO SELF · SIGHT ·
BITTER · DIGESTIVE
AND CIRCULATORY
SYSTEMS · FRUIT
AND SEEDS OF
PLANTS · SUMMER ·
ARIES, LEO,
SAGITTARIUS ·
WANDS · EAST

VISUALIZATION:
PLATONIC SOLIDS – WORKING WITH THE ELEMENTS

▲ Hold each of the Platonic Solids in your mind's eye, one by one. Notice which you are drawn to first, how they make you feel and sensations that arise.

▲ Within each one, address your relationship with the element it represents, using the wheel above and the table on page 47, and discern any imbalances that you feel exist within your system. For example, do you have excess fire, producing anger, or too much water, creating boundary issues and leaking energy? Bring to mind the corresponding solid and hold it in your field of awareness, until you feel a shift that brings you back into alignment.

▲ Take this practice out into nature to immerse yourself in the direct perception of each of the elements.

PYRAMID POWER MEDITATION

Pyro – fire, *amid* – in the middle of; pyramid – in the middle of fire. Use this meditation to burn away anything that is no longer serving you, to recharge and light up your creative fire.

▲ Sit cross-legged with a hand rested on each knee so that you form a pyramid with your body. Allow a moment for your nervous system to acclimatize and land into the space.

▲ Take six deep breaths in through your nose, drawing the breath down through your spine and up through your solar plexus and heart, out through your mouth.

▲ In your mind's eye, draw a square on the ground around you. This will form the base of your pyramid. As you do so, acknowledge the four directions north, south, east, west, and the four elements earth, air, fire and water.

▲ Set the intention that this square represents the anchor to your physical reality, which will serve to ground the higher frequencies that you are calling in. Breathe into it.

▲ Draw four triangles up from the base of your square to complete your pyramid (*see* page 38). Do this by drawing a line from each corner of the square to just above your crown. The apex where these points join represents aether.

▲ Take another six breaths as before. Visualize a channel of white light piercing the tip of the pyramid and flowing into the space that you have created. Remember that you are made of light. Allow it to wash over you and be absorbed into your organs and every cell in your body.

▲ Once you feel the alignment and expansion of this activation, visualize a circle within the square. Spin it first anticlockwise to discharge any stagnant or blocked energy, then spin it clockwise to charge your energy centres.

▲ Visualize a flame of purification before you. Build a pile of all the things that you want to release and put them in the fire. Feel the lightness of letting go and the recalibration of your vibrational field.

▲ Now invite the fire of creativity and passion to ignite this space. Call to mind creative projects that you wish to guide this energy towards.

Merkaba

One of the most powerful sacred geometric shapes is
the Merkaba, which means 'light, spirit, body'. In the
Kabbalah in Jewish mysticism it is called a 'chariot of light'.
Sacred geometry is a form of light language, the structure
of cosmic communication. When a channeller receives
information into their subtle energy body, or when we
receive insight and healing, it is through sacred geometry
– dense etheric light that enters the nervous system and
produces certainty, a knowing, within the receiver.

In terms of physics, this is due to electromagnetism.
Light is produced by the transformation of energy between
electricity and magnetism. Ancient Egyptians understood
the Merkaba as a primordial frequency of light, an
electromagnetic force field that weaves the energetic fabric
of our universe. From hieroglyphs, we can break down its
components to *mer* (light), *ka* (spirit) and *ba* (body). In
quantum terms, *mer* is the electrical current or sine wave
that seeds existence or the waves of the primordial sea.
Ka is magnetism, and the two spin together to create an
electromagnetic field that gives birth to *ba*, the breath
of life that ignites our physical reality.

Each one of us moves inside our own Merkaba as an
etheric structure that exists on the periphery of our auric
(electromagnetic) field. By setting your Merkaba in motion,
you are creating a vehicle for exploring the psyche, delving
into the subconscious and the shadow self. It creates an
etheric gateway for deep cellular rejuvenation and shamanic
journeying, particularly effective during lucid dream states
as a method of moving between realms of consciousness.

In silence, it lies dormant. A field of pure potential, it waits to be activated through conscious connection. Once it is set on its spin, its field invokes higher forms of awareness. In lower states of consciousness, our capacity to integrate this light into our cells is limited. When we raise our level of consciousness, we can hold a higher density of this light, which facilitates healing and spiritual evolution. By providing the foundations to anchor cosmic energy into the physical realm, the Merkaba acts as a manifestation matrix to align with your highest potential.

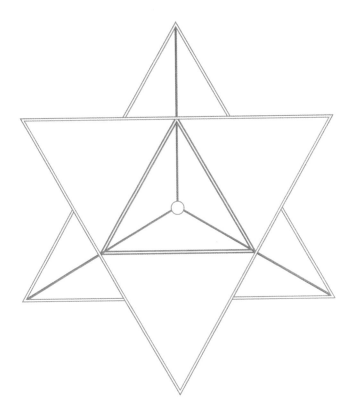

Merkaba – a star tetrahedron known in mysticism as a multi-dimensional vehicle of light.

MERKABA ACTIVATION

Activating your Merkaba is an extension of the Pyramid Power Meditation (*see* page 49).

▲ Once you have integrated the light of your pyramid into your body, create a mirror pyramid beneath you. Visualize its tip penetrating to the core of the Earth. This serves to balance the masculine and feminine energies within. The inverted pyramid resonates with the yielding, flowing, yin qualities of the feminine and the pyramid above is the active, creative, yang of the masculine. The base of the masculine pyramid is your knees/sacrum, and the feminine is your throat/shoulders.

▲ Feel into the kundalini energy. In Sanskrit, kundalini translates as 'coiled serpent' and refers to the powerful energy stored at the base of our spine that, when activated, releases a charge up to the crown of the head. Shakti – the primordial cosmic energy within all of us – is symbolized as a snake wrapped around a lingam stone.

▲ This is where it lies dormant for many of us. Visualize the helix, or double-headed serpent, spiralling up through your spine, vertebra by vertebra, until it reaches your crown. As it does this, your chakras (the energy centres of your body) open and release.

▲ Now invite the pyramids to spin; the top one turns clockwise and the bottom one anticlockwise. The movement of the two pyramids creates the star tetrahedron of the Merkaba. As it spins, it expands to approximately seven times the size of your physical body, filling your etheric body with light.

▲ You may now programme your Merkaba. What is it that you desire? Where would you like your Merkaba to take you?

Number

'I prove a theorem and the house expands:
the windows jerk free to hover near the ceiling,
the ceiling floats away with a sigh.'

Geometry by Rita Dove, former US Poet Laureate (b.1952)

Now that we have explored the birth of our universe
through a cosmological and mythological lens, we begin
our quest through the numerical ages. If geometry is a
language, numbers are its vowels and consonants. Numbers
are storytellers, revealing universal patterns and providing
clues to the mysteries of life.

Once considered a magical art to be practiced by initiates, mathematics not only helps us to understand our world, it also allows us to explore the parts that are beyond our reach – distant galaxies, black holes and the furthest corners of our imagination. Equations enable us to predict possible realities and develop as a species through scientific discovery.

Greek mathematician and philosopher Pythagoras (*c*.570– 495 BC) saw whole numbers as sacred entities in their own right – magicians of manifestation, each with their own personalities, mannerisms and tricks, each in a relationship with the other. Numbers can be mysterious or brazen, positive or negative. They can get on with other numbers or they can clash; they can create or they can destroy.

RIDDLE OF THE SPHINX

In Greek mythology, the Sphinx – with the head of a woman, the body of a lion and the wings of a bird – was a demonic creature that plagued the city of Thebes. She demanded a solution to the riddle – *What animal is it that in the morning goes on four feet, at noon on two feet, and in the evening on three?* – and destroyed those that answered incorrectly.

Oedipus answered correctly – man. A baby crawls on all fours, an adult stands on two legs and in old age, a person may need the support of a stick.

Occultist Manly P Hall (1901–1990) suggests that in this riddle we can see the journey of mankind. The baby on all fours signifies the innocent fool, earthbound to the physical and material world. Two is duality and intellect and three is the transcendence over the mental plane to spiritual wisdom and evolution. In this way, the Sphinx holds the key to the mysteries of nature.

The tarot

The archetypal quest of the human psyche is illustrated in the way of the tarot and we will explore this as a way of relating numbers to experience along our journey. Alejandro Jodorowsky (b.1929), surrealist director and shamanic psychotherapist, calls the tarot 'an architecture of the soul'.

The tarot is a pack of 78 playing cards, reportedly dating back to 15th-century France, and is used as a divination tool to shine light on the life of the questioner or player, known as the querent. The pack is divided into arcanas, or secrets. The minor arcana comprises 56 cards that fall into four suits:

▲ coins or pentacles (earth)
▲ cups (water)
▲ swords (air)
▲ wands (fire)

The major arcana has 22 cards, which are not suited but heavily symbolic and archetypal. They represent the Fool's journey from a material state of childlike innocence, through the trials and tribulations of life's process of discovery, overcoming the mental state and, through wisdom and experience, reaching the spiritual realm and the spiritualization of matter, returning once more to the Fool.

Here we will be focusing on the first 10 major arcana. Numbers 1 to 10 can be drawn as sacred forms nested into one another, as shown on the front cover of this book. The Meditative Geometry chapter (*see* page 143) offers instructions on how to draw each shape. You may like to follow them as we go through the shapes here.

0/1/MONAD/DOT-LINE-CIRCLE

Ancient numerical systems began with
the number one. Zero did not find roots
in Europe until around the 12th century. The Sanskrit for zero
is *sunya*, meaning empty, but referring to the pregnant
cosmic void that is full of potential. In the West, this was
considered a dangerous and chaotic concept because there
could never be nothing – there had to be God.

Monad, the number one, represents the origins of being, the
singularity. One is perfect, whole and complete. Initiating
and pioneering, it is the first bubble of the primordial soup,
the creative seed that gives birth to all other numbers:

$$111111111 \times 111111111 = 12345678987654321$$

It never creates duality; it is always in absolute unity.
If we multiply or divide a number by 1, the number
remains unchanged. For example, $1 \times 9 = 9$; $3 \div 1 = 3$.

The binary code that we use for computer technology
consists of just 0s and 1s. Binary means to have two parts.
All possibilities and all other numbers are created from
this 'oneness'. Gottfried Wilhelm Leibniz (1646–1716), the
17th-century philosopher and mathematician who created
the binary code, said he was inspired by the *I Ching*, also
known as the *Book of Changes*. This is a sacred divination
tool that uses yarrow sticks to create 64 oracles in the form
of hexagrams, the combinations of which all come from the
binary roots of yin and yang. The upper half represents the
heavens and the lower half, the Earth.

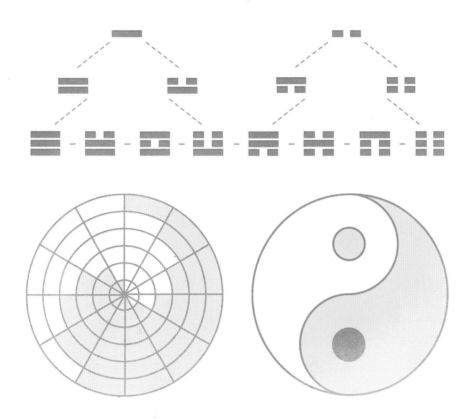

Creation: the *I Ching* (top),
binary wheel (bottom left)
and yin–yang (bottom right).

The Fool and the Magician

In the tarot deck, the unnumbered card, sometimes counted as zero, is the Fool. As all beginnings come from endings and endings lead to beginnings, it can be placed at the beginning or end of the pack. The Fool is full of enthusiasm and naivety as he embarks on the next phase of his journey. Number one is the Magician, who has boundless creative potential ahead and faces the challenge of bringing spirit into matter. We take our first step.

2/DUAD/VESICA PISCIS

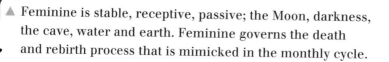

Two is the first even number. The Ancient Greeks decided that even numbers were feminine and odd numbers masculine – giving numbers a gender reflects the opposing and complementary creative qualities within all of us. We are each a blend of both principles.

▲ Feminine is stable, receptive, passive; the Moon, darkness, the cave, water and earth. Feminine governs the death and rebirth process that is mimicked in the monthly cycle.

▲ Masculine is unstable, active, outwards, forceful; the Sun, fire and air. It is generative and igniting.

Two is also the first number that can be multiplied by itself to produce a greater number: $1 \times 1 = 1$ but $2 \times 2 = 4$. It can create and bring about new forms of existence. It is about the coming together of two things and the exchange of ideas or lives; attraction and repulsion; the relationship of self to others. Through confronting the duality within us we are able to learn and transform.

The High Priestess

The second card of the major arcana, the High Priestess, adapts and assimilates. She weighs up the knowledge around her, digests her experiences and considers her next move.

▲ We have two eyes, two hands, two feet, which give us balance and dexterity.

▲ Our DNA bases form pairs.

3/TRIAD/TRIANGLE, TETRAHEDRON

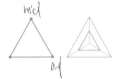

Three is an eruption of creative energy leading to resolution. The marriage of one and two. We feel a sense of completion when things are done in threes – beginning, middle, end.

This masculine number is an active and expressive force. The triangle is the alchemical symbol for fire and symbolizes its creative and destructive powers. Pythagoras held that this was the ignition of creation and the first 'true' number.

The Empress

The third major arcana card represents desire, impulse, fertility, reproductive forces and action towards manifesting in the world.

▲ The three primary colours of red, yellow and blue make up our palate.

▲ We have 33 vertebrae in our spine with 33 pairs of nerves.

4/TETRAD/SQUARE, CUBE

Four is physical manifestation. It stabilizes the fiery urges of three and lands us in the material world. It is represented in the cube of the earth element in the Platonic Solids (*see* page 46). An even number, it is considered feminine and in many ancient cultures signified Mother Earth and her many goddess manifestations. It is also the human physical form that has materialized out of the union of the trinity. It is the four directions of the Earth, the four seasons and what Jung calls the four natures of the self – sense, emotion, thought and intuition. It is the four forces of nature – electromagnetism, gravity, strong and weak nuclear forces.

The Emperor

The fourth card is the Emperor, stable and fixed in the material world, watching over his domain.

5/PENTAD/PENTAGON, PENTAGRAM

Five is the aether element. Moving out of the security of four, it propels us on our quest and acts as a bridge to other worlds. The last of the Platonic Solids (*see* page 46), it represents spirit, transformation and liberation. In alchemy, it is the quintessence – one to four refined and distilled to their most exalted state. It is sensual, abundant and the signature of nature.

Many trees and shrubs that bear edible fruit have five-petal flowers. Cut an apple in half at its equator and you will see a perfect example of this. Pentagonal symmetry is everywhere in nature: starfish, maple leaf, starfruit, seeds, paws, five fingers of a hand. It signals strength, power and fertility. Five introduces the star and communicates the relationship between the cosmos and earthly realms.

The pentagon is drawn in one free-flowing motion, honouring each of the elements and welcoming abundance, attraction and protection. This five-pointed star was the emblem of the Pythagoreans.

Free-flowing pentagram (top); Pentagram and the elements (bottom).

'Man is disposed according to the number five: he is of five equal parts in height and five in girth; he has five senses and five members, echoed in the hand as five fingers.'

·Hildegard von Bingen (1098–1179), philosopher and polymath

The Hierophant

The fifth card in the major arcana, the Hierophant, is also known as the Pope. Moving away from the rigidity of the Emperor yet still concerned with structure and order, he is concerned with ecological over economical harmony and the spirit of his people.

72

- ▲ This is a potent number in sacred mythology.
- ▲ According to Vedic wisdom, there are 72,000 *nadis* (life force channels within the body).
- ▲ There are 72 names for God in the Jewish Kabbalah.

- ▲ Legend has it that 72 languages were spoken in the Tower of Babel.
- ▲ The exterior angle of the pentagon is 72 degrees. The interior angle is 108 degrees (*see* page 82).

6/HEXAD/HEXAGON, HEXAGRAM

The number six is about pleasure, love and perfection through balance. In the Tree of Life, the sixth sephirah is Tiphareth, signifying beauty. Now we are able to witness the fruits of existence. As we saw in the creation myths, this is the ultimate day of creation. Represented by the hexagon, the number six is also found in the Flower of Life.

Six is the union of the two equal triangles that are born from the Vesica Piscis, representing the masculine and feminine principles brought into harmony. Pythagoras viewed the birth of numbers to start with number three; six is the summation of that initial creation: $1 + 2 + 3 = 6$.

Hexagons are prolific in nature and are in charge of function, structure and efficiency. Honeycombs and tortoise shells, the skins of fruits and animals, they are even formed by the friction of molecules in liquid. All snowflakes, though unique in design, follow a hexagonal pattern, forming around an angle of 60 degrees. This and the hexagonal arrangements of hydrogen and oxygen bonds of a water molecule has led to six being the symbolic number for water, flow and abundance.

THE SEAL OF SOLOMON

The hexagram star, also called the seal of Solomon, is named after King Solomon (970–931 BC), thought to be a master magician who derived his powers from a ring engraved with its seal. In occult tradition, this 'creator's star' represents the human soul, the equilibrium of the masculine principle of fire (upward triangle) and feminine principle of water (downward triangle). It symbolizes the alchemical doctrine 'as above, so below' – a marriage of the heavenly and earthly realms.

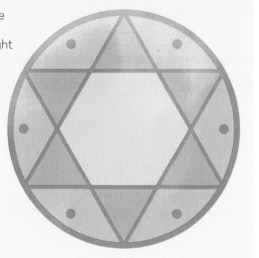

The Lovers

The sixth card of the major arcana, the Lovers celebrates ultimate union, harmony and pleasure.

12

- ▲ Twelve, twice six, is the number of signs in the zodiac, and the number of hours on a clock. It represents our movement through time and the cosmic order. Time is measured in sixes and 12s: 60 seconds, 60 minutes, 12 sections on a clock and 24 (2 + 4 = 6) hours in a day.
- ▲ There are 12 tissue salts in the human body.

- ▲ Twelve is the measure of the hero's journey: there are 12 ordeals of Odysseus and 12 labours of Hercules.
- ▲ Jesus had 12 disciples.
- ▲ The dodecahedron has 12 pentagonal faces and is revered as the shape of transformation and regeneration.

7/HEPTAD/HEPTAGON

Seven is considered to be the sacred number of the world as it is the number of days of creation. According to Rudolf Steiner's theory based on planetary influences, our lives are cast in seven-year cycles of evolution – a shift occurs in the seventh year and then comes eight, the end of one cycle and the beginning of another. At age seven, the individual personality and desire for independence becomes clearly developed and this is often signified by a new set of teeth. By 14, puberty has happened and adolescence is occurring. At 21 we reach full adulthood and our teeth stop growing.

- There are seven days of the week, once named after the seven visible celestial bodies in the sky.
- The rainbow has seven colours.

- There are seven main chakras and seven endocrine glands.
- Pythagoras called number seven the 'vehicle of life'.

The Chariot
The seventh card in the major arcana, the Chariot reflects the individual's journey out into the world.

8/OGDAD/OCTAGON, OCTAHEDRON

Eight is a pinnacle point in our journey. We have reached a state of maximum experience and an understanding of the cycles of life and that death and resurrection are part of it. It is the goal and once it is attained, it heralds a new beginning. Turned on its side it is the infinity symbol or lemniscate, representing the never-ending cycle of life.

VISUALIZATION: ADJUST YOUR BALANCE

▲ Trace an energetic lemniscate or figure of eight running vertically on your body so that the top loop goes around your head and torso, the bottom loop goes around your feet, with its axis crossing over your solar plexus.

▲ Begin to spin it so that it is synchronizing both parts of your body, your north and south poles, upper and lower chakras as one.

▲ Allow the energies to flow back and forth.

▲ Do this until you feel equilibrium.

▲ Spin the lemniscate on its side so that it is horizontal, with its axis over your heart.

▲ Repeat the same process, entraining the left and right hemispheres of your brain and the left (feminine) and right (masculine) sides of your body.

Justice

The eighth card in the major arcana is represented by the scales of life. We have reached the point of judgement, recognition and evaluation.

8

▲ The number eight is considered deeply transformational in esoteric texts. In Buddhism an eightfold noble path leads to enlightenment. In Egyptian mythology, there are eight primordial gods. Jung identified eight cognitive functions – extroverted and introverted expressions of thinking, feeling, sensation and intuition.

▲ Eight symbolizes wealth and prosperity. The Hindu goddess Lakshmi, who represents those qualities, can inhabit eight different forms.

▲ Eight is the number for oxygen in the periodic table of elements and the octahedron represents the air element in the Platonic Solids (see page 46).

▲ $8 \times 8 = 64$ (see below).

64

▲ Throughout cultural history 64 is particularly significant. There are 64 hexagrams in the Chinese *I Ching*; in Vedic texts, 64 books of the Indian tantras; 64 sexual positions in the *Kama Sutra*; and 64 taste combinations in the Ayurveda.

▲ Squares on a chessboard number 64. Chess pieces have a significant relevance to the psychological quest of the soul and their symbolism is encrypted into the esoteric practices of occultist groups, such as the Freemasons and Rosicrucians.

▲ There are 64 possible patterns of codons (specific sequences of three) in human DNA.

9/ENNEAD/NONAGON

Pythagoras thought of nine as the protector and guardian of the other numbers, a cosmic gatekeeper that holds everything together, keeping potential chaos at bay.

The Hermit

The ninth major arcana card presents a crossroads. Are we prepared to sacrifice everything that we have accumulated on our journey so far and ready for another quest? The card is the Hermit, who retreats and reviews the experiences he has had and the lessons that have unfolded. He listens deeply to his inner guidance as he transforms from the ego-self to the universal self.

▲ When nine is multiplied by any other number, it always reduces back to nine: $3 \times 9 = 27$; $27 (2 + 7) = 9$. $7 \times 9 = 63$; $63 (6 + 3) = 9$.

▲ Celtic God Odin hung for nine days from the Yggdrasil tree that connects to the nine worlds.

▲ Human gestation takes an average of nine months.

▲ There are nine muses of Ancient Greek mythology.

▲ In Aztec and Mayan mythology there are nine earthly layers with nine gods.

10/DECAD/DECAGON

Ten is the end of our journey. It is the setting sun and we wait for a new beginning to arise.

$1 + 0 = 1$: we are at the end and the beginning. Following on from nine, ten contains the zero and one of our beginning and all the numbers of our travels. Now the future is made out of all the numbers contained within it.

The Wheel of Fortune

The tenth card in the major arcana represents the cyclical nature of life and, in Eastern philosophy, the karmic process whereby souls reincarnate into lives based on conditions fruiting from previous lifetimes. It signifies the intersection between planes of reality, where the bandwidth in our field of experience expands and contracts, depending on our actions so far.

PYTHAGOREAN TETRACTYS

The tetractys ('decade') consists of the numbers 1 to 4 layered together to form a 10-point triangle. It signifies the journey of entering and exiting human life. Incarnating from (1) spirit, it descends through (2) mental, (3) emotional and finally (4) physical realms. Through spiritual enlightenment or dying, we reverse the process, transcending the physical into spirit. Pythagoras so revered his creation that his initiates had to swear an oath on it. It is possible to read the tarot using the tetractys as a spread, laying out the cards to provide you with the spiritual, mental, emotional and physical responses to your question as well as the path of progression from past to present and future.

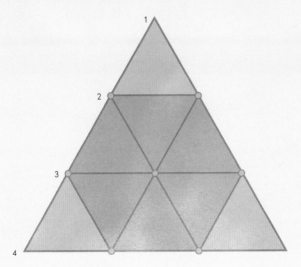

Sound

'Before its incarnation the soul is sound. It is for this reason that we love sound.'

Hazrat Inayat Khan (1882–1927), Sufi master

In 1952, composer John Cage (1912–1992) wrote a silent composition called 4'33", which was inspired by a visit to an anechoic (echoless) chamber at Harvard University. Expecting to be deafened by silence, he was instead met with a range of sounds coming from inside his body. He could hear the popping and crackling of his nervous system and the swoosh of fluid in his glands, the beat of his heart and the pulse in his neck, a high-pitched ringing in his ears. Our biochemistry is a symphony with trillions of cells making up the orchestra. The natural world is always performing, from the jubilant birdsong of the dawn chorus to the eerie whistle of desert sand dunes and deep ocean-bed murmurs.

Interference patterns produced by the meeting of sound waves, demonstrating the Vesica Piscis in dynamic motion.

The Ancient Greeks perceived the harmonics arising in the cosmos as an orchestral dance, calling it *musica universalis.* Pythagoras theorized that each of the planets had its own frequency or cosmic hum based on its unique orbit, and called this the harmony of the spheres. In recent times, NASA spacecraft have recorded the radio waves emitted by the planets, which have been converted into haunting audio.

Harmonics of a vibrating string

Pythagoras came up with the idea of applying mathematics
to sound after hearing the different notes that a blacksmith's
hammers made when struck. He figured out that this
was due to the different weight ratios of the hammers.
A hammer weighing half as much as its neighbour sounded
twice as high a note – a ratio of 2:1. This is what we call an
octave. Pythagoras found that plucking a string at certain
points made pleasant sounds while plucking it in other
places resulted in sounds that were quite jarring. He found
that only whole numbers produced a harmonious sound.
A ratio of 3:2 was particularly sweet on the ear and became
known as the 'perfect fifth'. This relates to the Golden Ratio,
written using the Greek
letter φ (*phi*). Pythagoras
created a scale based on this
and the octave. Examples
in classical music include
the first movement of
Beethoven's *Fifth Symphony*
and Debussy's *Dialogue du
Vent et la Mer*. The Golden
Ratio, also known as the
divine proportion, is a
fundamental growth pattern
in nature which we will
explore in the next chapter,
on Nature (*see* page 90).

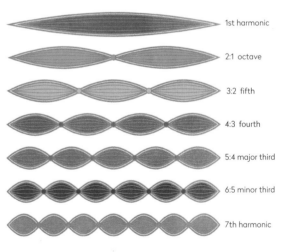

1st harmonic

2:1 octave

3:2 fifth

4:3 fourth

5:4 major third

6:5 minor third

7th harmonic

The harmonic scale, from
which the Pythagorean
scale evolved using the
octave (2:1) and perfect
fifth (3:2).

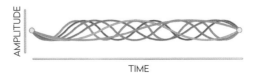

AMPLITUDE

TIME

Sacred sound

Music has always been a significant tool for ceremony and contemplation. Spiritual music has a twofold purpose. Prayers and intention float on the airwaves through music and up into the heavens. It is intended to invite beings from the celestial realms down to Earth and also to raise the spirits and awareness of humans up to the celestial realms, allowing people to come into resonance with higher spiritual principles, raising their own vibration and level of consciousness.

LET THERE BE MUSIC

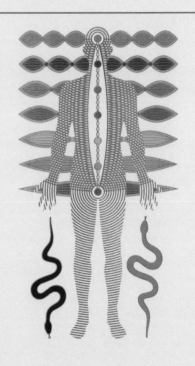

- ▲ Apollo, the Greek god of music, would play his sacred lyre as a form of enchantment and to promote healing.
- ▲ In Sufism, music, chanting and ecstatic dance are the means used to achieve altered states of consciousness to bring the devotee closer to original divine source.
- ▲ The primal, instinctual power of healing through the sound of the human voice, from singing a lullaby to intoning chakras, has been an intrinsic part of our cultural evolution.

Sound healing

This ancient practice is evident in the harmonic resonance chambers of the pyramids where it is thought that initiates would attune their vibration to each chamber as they raised their frequency, opening up new channels of information and higher levels of cosmic understanding. According to Egyptian scholar and indigenous wisdom keeper Abd'el Hakim Awyan (1926–2008), patients would lie on the stone slabs of the healing temples at Saqqara, Egypt, and the deified physician Imhotep and his successors would sit in another chamber from where they could listen to the sound of the patients as their vibrations channelled through the stone. From this, they would be able to diagnose where there was an energetic imbalance in the body which was creating a physical disturbance.

We are beings of resonance and in constant vibratory exchange with our environment. Sacred vibrations and instruments such as tuning forks, singing bowls and shamanic drums are used to shift stagnant vibratory patterns causing dis-ease. To bathe in sound is to allow every cell in your body to be washed with its vibration and in doing so, attune your frequency to resonate at a desired level.

LISTEN TO YOURSELF

Be still, cover your ears and see what, like Cage in his anechoic chamber, you can hear. In deep states of meditation, it is possible to get a sense of the pace at which your blood is flowing, the air pockets cracking in your joints and the overall tone at which your body is vibrating. Listen to the sound of your silence.

BRAIN WAVES AND STATES OF MEDITATION

▲ **Gamma** (40Hz +): insight, eureka moments, channelling.

▲ **Beta** (14–40Hz): normal waking consciousness, intelligent thinking, anxiety and stress.

▲ **Alpha** (7.5–14Hz): deep relaxation, serotonin release, left and right brain hemisphere processing.

▲ **Theta** (4–7.5/8Hz): meditation, transcendental/altered consciousness, shamanic journeying, plant medicine, ceremony, sleeping.

▲ **Delta** (0.5–4Hz): deep sleep.

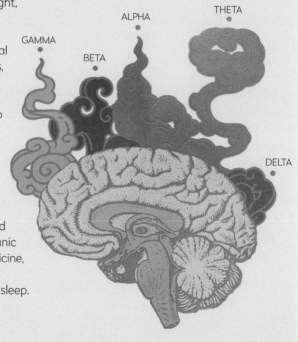

Brain waves

Like sound, brain-wave states are measured in hertz (Hz), which indicate the number of cycles in a second. Earth's own electromagnetic frequency, the Schumann resonance, averages at 7.83Hz, which is on the border of our theta and alpha states. When we do not resonate with this frequency, our bodies begin to weaken and become vulnerable to disease. For this reason, NASA created Schumann resonance generators in their spacecraft. Since the brain is a finely calibrated electromagnetic organ, any fluctuations in the geomagnetic activity of the Earth that alter this resonance, such as solar flares and lightning, also alter brain and neuro-hormone responses.

'Our entire biological system – the brain and the Earth itself – work on the same frequencies.'

Nikola Tesla (1856–1943), engineer and inventor

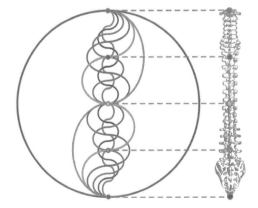

The body as an instrument: the rise of kundalini energy up the spine.

Sound and vibration can alter our biochemistry and, in doing so, our physiological wellbeing, causing neurotransmitters such as dopamine and serotonin to flood the brain. A fast tempo can trigger the release of adrenaline; hypnotic tones can produce its opposite, noradrenaline. Our brain waves are altered, our mood shifts and every organ, cell and micro-organism within us responds to the call. In Western medicine, sound waves are used to break up kidney stones.

When two frequencies meet and their cycles align, they create a harmonic resonance. By bringing sacred sound and shapes into our sphere of awareness, we are tuning in to their vibration. Low vibrational states include stress, disease and negative emotional patterns. By raising our frequency to align with higher rates of consciousness we widen our field of awareness and the depth of our experience.

Cymatics

Coined by Hans Jenny (1904–1972) in his book of the same title, the term cymatics refers to the demonstration of how sound waves produce geometric patterns. Chladni plates, created by Ernst Chladni in the 18th century, are connected to a machine that plays a tone. On top of the plates are grains of lycopodium powder or fine grains of sand, which come together to form shapes. The higher the frequency, the more complex the patterns produced.

Vibrational pattern produced by the sound 'Om'.

▲ Jenny found that the pronunciation of Sanskrit and Hebrew vowels produced shapes that matched their written symbols.

▲ In Hindu practice, a mantra is a sacred sound and repeating it out loud creates a geometric pattern, through which a call is sent to the universe. The power of intention affirms the desire of the mantra and amplifies its vibration. Mantras are often repeated 108 times.

▲ An occult number, 108 is held sacred in many religions: it is the number of beads in a prayer mala (prayer beads) in Hindu and Buddhist religions; Hindu deities have 108 names; Yogic sun salutations are performed in nine rounds of twelve (108); there are 108 marma points (junctions of the *nadi* channels – *see* page 64) in Vedic tradition – 107 in the body and another one in the mind.

▲ The diameter of the Moon is approximately 2,160 miles, corresponding to the length of each zodiacal age. Its radius is 1080 (half the diameter).

▲ The ancients associated the Moon energetically with silver, which we now know has an atomic weight of 108. If we multiply 2,160 by 108 we get 233,280, which is about the distance in miles between the Earth and the Moon.

Sound vibration and water memory

Japanese researcher Masaru Emoto (1943–2014) used a magnetic resonance analyzer to photograph the structure of water under a variety of conditions to demonstrate how it responds to the electromagnetic field of its environment. In his book *The Hidden Messages in Water*, he documents the crystalline shapes of the molecules produced when exposed to a variety of conditions. Distortion occurs under unfavourable conditions, such as local toxins, geopathic stress, electromagnetic pollution and negative emotional states. Pure and harmonious conditions, such as unadulterated nature, sacred space and positive emotional states, yield balanced and aesthetically appealing shapes. The term he developed for the sacred geometry revealed in his findings is the hado principle (hado translates as 'vibration'). He experimented with the impact of vibrational healing and sound through which he found he was able to alter structure and transform distortion into beauty.

Water memory is used in vibrational medicines, such as flower and crystal essences and homeopathy, to transmit the energetic signature – the sacred geometric structure – of a subject (plant, mineral, animal, and so on) into a patient. The body, fluent in the language of vibration, recognizes this pattern and restructures itself accordingly.

Nature

'Come forth into the light of things. Let nature be your teacher.'

William Wordsworth (1770–1850), English poet

Nature is in a constant state of dialogue, an evolutionary metamorphosis that is love in its most primal form – attraction and bonding, destruction and renewal, giving and receiving. This communication is through the energetic nervous system of sacred geometry, rewriting the structure of its existence through pattern and form.

In alchemical and shamanic traditions, this is called the 'green tongue' or the 'language of the birds', a divine holographic and symbolic communication channelled between the initiated and beings of elemental and spiritual realms. Swiss alchemist and physician Paracelsus (*c*.1493– 1541) called this web of exchange 'the light of nature'. In many traditions, birds act as chief messengers between worlds, passing information to animals, plants, insects and elemental beings through their song. To be fluent in this language is to unlock the mysteries and magic of the universe.

Listen to nature

Through the ages, so much has been lost in translation. Since the Industrial Revolution our ability to listen to nature has waned and we must relearn its language, how to be in communion with the natural world and allow it to be our teacher and guide. As an apprentice, co-creating with nature, we open ourselves up to receive wisdom that has been passed down through the library of evolution. Philosopher and theologian Jacob Böhme (1575–1624) identified this divine intelligence of nature as *natura sophia*, after Sophia the Gnostic goddess of wisdom.

Nature courts many lovers, inducing a symbiotic relationship of call and response. The beat of its heart and the wisdom of its soul are felt by all of its subjects and it is never not there, never not inviting you into its embrace. When we are immersed within it, our hearts not our minds receive its wisdom. Everything in nature, including ourselves, is a microcosmic expression of the macrocosm.

As above, so below

Asteroids corkscrew through space, moons orbit their
celestial bodies and planets swing on their axes and their
dusty skirts twirl like whirling dervishes. This cosmic
dance traces complex geometry, recording every proportion
and interaction created by these celestial spheres moving
in relation to one another. These designs echo the patterns
of nature seen on Earth and provide the blueprint for
sacred design.

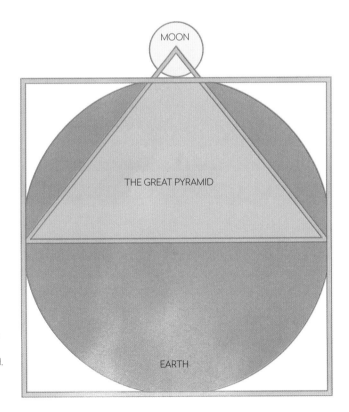

Planetary proportions:
the Earth and Moon in
relation to each other
and the Great Pyramid.

The dance of Venus

If we observe the orbit of Venus around the Sun, from the perspective of the Earth, in every eight Earth years (13 Venutian years) there are five loops – retrograde cycles – when Venus appears to stand still in the sky, 'kissing Earth', at each point creating the tip of a petal that makes up a five-petalled rose, the Rose of Venus. In nature, we see this pattern repeated in the pentagonal seed pattern of an apple and the form of a starfish.

The orbital path of the planet Venus forming a five-petalled rose, based on the work of Hartmut Warm.

LIVING DESIGN

The living spiral of a nautilus shell or the horn of a sheep, the interlocking hexagons of a beehive, the underground formation of crystals, the spin of a spider's web and the formations of migrating birds are designs that go beyond the aesthetic. Their beauty is functional. Their structure creates the specific dynamic, strength and balance required to support the role of the individual and collective.

Golden proportions: *phi*
and the Golden Spiral.

Golden Spiral

This is the Golden Ratio (expressed as φ or *phi*, *see* page 75) in practice. It begins with a series of nesting 'golden rectangles' that spiral out, using the *phi* ratio, into 'golden squares', maintaining the shape as it grows. This logarithmic spiral provides the basic curves of growth patterns and governs natural cycles and can be found in your body, the food you eat, the plants you grow and everything in between. It reaches far beyond our earthly realm to rule the planets.

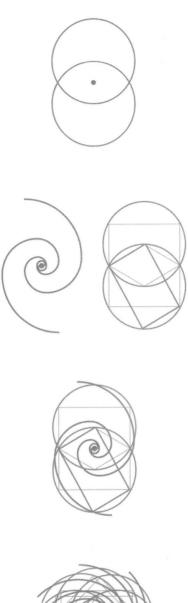

Vesica Piscis and
the Golden Ratio.

Fibonacci sequence

This numerical sequence describes growth patterns, such as cell division, in living structures. Each number is the sum of the two previous numbers: 0, 1, 2, 3, 5, 8, 13, 21, 34, 55, 89, 144 and so on. The sequence is closely associated with the Golden Ratio since the ratio of each successive pair of numbers is nearly 1.618 or *phi*. But as *phi* has no end, it is impossible to attain, only to get very, very, very close. It signifies striving towards perfection and journeying towards a divine source.

The Fibonacci sequence determines the number of petals on a flower, which tend to be in sequences of 5, 8 and 13, but can also be 3, 21, 34, 55 and so on. Two Fibonacci spirals occur regularly in nature, such as the clockwise and anti-clockwise spirals of a sunflower, growing in opposite directions to create a system of complete balance and stability during the organic growth process, like stabilizers on a bicycle. Think of a young seedling winding its leaves around its stem. This even distribution of leaves also allows for maximum exposure to water and sunlight.

Fibonacci growth patterns: the clockwise and anti-clockwise distribution of petals on a sunflower head (left) and the spiral formation of leaves around the stem of a plant (right).

THE HOLY GRAIL OF NUMBERS

The numerical value of *phi*, used to express the Golden Ratio, is usually simplified to 1.618 but really it is more like: 1.6180339887498948482045868343 65 638117720309... and on to infinity. It is known as an irrational number because it is not whole and can never be pinned down to a simple fraction – a marvel of beauty and perfection.

LOOK AND SEE

Observe the patterns of nature in fruits and vegetables by cutting them in half. Notice the Golden Spiral in a red cabbage and the five-pointed star of an apple.

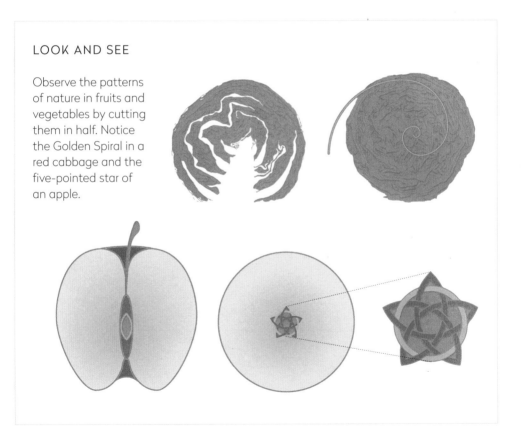

The holographic universe and the mirror that nature holds up to us is at play in a further layer of ecological linguistics that is known as the doctrine of signatures (sign-nature: signs in nature). The proposition is that nature provides visual clues so that we may know how to work with plants for healing and for what purposes. For example, walnuts resemble the human brain and support cognitive function due to high levels of DHA (omega-3 fatty acid).

The fractal geometry of nature

Fractals are complex, self-replicating shapes that look more or less the same on a sliding scale. Popularized by mathematician Benoit Mandelbrot (1924–2010), nature is loaded with fractals – the repetition of pattern throughout organic life. We can witness this in how the branching of a tree is the same as the tributary of a river or the bronchioles of our lungs. When we observe a sea-churned and limpet-covered rock on the shore it bears striking similarity to the surface of the Moon.

Whether you are looking at something from outer space or with a magnifying glass, it will hold its essential pattern (known as self-similarity). One way to consider the energetics of sacred geometry is that, if you continued to magnify an object to beyond the subatomic level, beyond all matter, then you would be left only with its essence, its energetic blueprint.

FRACTAL PROGRESSION
The shape of the pattern is the framework, that is the part that makes a leopard a leopard. What makes the difference within that framework is our particular energetic signature for life. The way that each leopard is unique is in its personal splendour, the colour of its coat, the round of its eye and the irregularity of its ear. The pattern replicates based on the quality of the vibration that came before, which is in a constant state of flux. This is known as fractal progression. This is how we can transform our physical selves through energetics, by shifting our patterns and consciousness. Then the next version of us produced will be stronger, more vital than the one before.

Koch snowflake variations – one of the earliest documentations of fractal progression.

Walk the ... look for a definitive boundar... water. From a distance ... oser and from your hands and knees see the jutting rocks, the undulations of land, the rough composite of minerals that have built up and eroded over the years, battling the storms of evolution. Think about how definitive your conclusions might be if you were a beetle or an ant and how that perspective shifts again if we go to a microscopic level.

Coherence and growth

Sacred geometric patterns are archetypal, so they provide the fertile soil from which the fractal progression of our growth can spring. We are working with these shapes to refine and bring the patterns into coherence in the body and our environment, the spaces that we occupy and our interactions with people. The conditions of this soil determine the quality and structure of the blueprints for the future. Every moment has the potential to fertilize the ground for renewal and transformation. This is the key to deep healing through the practice of sacred geometry in a fractal world.

▲ Based on what you know about sacred geometry so far, choose a shape that particularly resonates with you, or contains a quality with which you would like to align.

▲ Take a few deep breaths in to your heart and out via your solar plexus to regulate your system into a receptive state.

▲ Now, invite the shape into your field of awareness.

▲ Visualize it as a hologram, replicating into infinite versions of itself, a wave of imprints that merge with your consciousness.

▲ Imagine these shapes landing on every cell in your body, anchoring and locking into the cell, and in doing so, transforming the energetic blueprint of the cell into a coherent, new emergent state.

Natural co-existence

[Nothing is in isolation; everything functions as part of a system.]In his systems theory, physicist Fritjof Capra (b.1939) believes that, if we take our lead from nature, we can learn to move away from reductionist, individualized patterns in society towards radical new frameworks for rewriting our economic, political and social systems. Much of modern science is concerned with experimentation through separation and isolation, yet this does not exist in nature, everything is part of the collective and in order to work with the magic of a plant it is necessary to go beyond the sum of its parts. Within the whole, there is a complex metamorphic system that is the poison and the cure, the light and the dark. This is the wisdom that comes from systems thinking.

It is vital for the future of our planet that we go back to the ways of our ancestors and co-exist with nature once again. We have borrowed so much from nature in our creations – our mechanics, engineering, technology, the fabric of our physical world. Just about every facet of our manmade reality has been sourced from the design of nature, and everything is powered by it. There is so much more to learn. Nature can help us to undo our wrongs and return to balance, but we must work with it, not against it. Sacred geometry is a systems language that provides an evolutionary framework. It allows the dots to be joined and new patterns and properties to emerge.

DNA COMMUNICATION

We are designed to communicate with the natural world and its deep intelligence. Indigenous shamans communicate through direct perception, using the consciousness of their heart and subtle bodies to enter into a dialogue with plants. Anthropologist Dr Jeremy Narby, author of *The Cosmic Serpent: DNA and the Origins of Knowledge*, spent time with Ashaninca people in the Peruvian Amazon. When he asked how they knew from which plants to make medicine for a particular disease, he was told 'the plants tell us'. He embarked on plant spirit journeys with the shamans, working with sacred plants, such as ayahuasca, and had visions of the double-headed serpent, the caduceus of healing that resembles the double helix of DNA. He developed a theory that, in an altered state of consciousness, our DNA is able to communicate directly with the genetics of the plants.

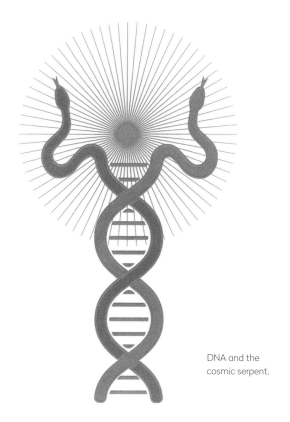

DNA and the cosmic serpent.

When we observe this continuum of life, when we fully turn up to the experience, we witness its perfection in its patterns of expression. The more we merge ourselves with the natural world, the more we see ourselves in its reflection. At first, we may see what appears to be chaos, but when we observe closely, we see a pure, complex, intricate system of design. There is intelligent self-organization at work. Electrons dance around atoms, holding the hands of potential partners and twirling together to form new groups; and new ways of being emerge, spiraling through time and space. A stem twirls around a branch, an egg falls from an abandoned nest, a seed head explodes and two antlers clash in a dual. Everything is in motion: adapting, evolving, resolving.

Plant communication

Behind every disease pattern is an energetic signal, a pulse
that is calling out, a rupture, a crack in the fabric of our
being, and it is this that the plant is responding to. By tuning
in to the energetics of plants and aligning our vibration with
theirs, we can, through sympathetic resonance, receive their
healing and wisdom. Sit under a tree and it will shower you
with its sacred geometry. When we open our hearts, we are
able to decode the information that exists within nature
if we allow plants to 'talk' to us, to divulge their ancient
wisdom. Our energy fields communicate with everything
that we come into contact with and by immersing ourselves
in nature, we allow it to become the architect of our own
energetics. It begins to redesign our geometry, activate the
parasympathetic nervous system, rewire neural pathways
and upgrade the crystalline structure of cellular membranes
– it recalibrates our antennae.

When we are out in nature, sitting before a plant, we
become the essence of the plant that we are working with,
the water of our bodies mirrors its vibration and our sacred
geometry is one.

PLANT COMMUNICATION EXERCISE

▲ Go out in nature. Make sure to turn your phone off as it can distort or block communication with your surroundings.

▲ Find a place where you can be alone, without distractions, and turn down the volume on your thoughts.

▲ Allow your intuition to guide you towards a particular plant. Sit next to it and allow your body to relax.

▲ There is nothing to force, just breathe and be present. Turn up fully into the moment. This is an exercise in feeling from your heart and communicating on a cellular level.

▲ You may sense a point when you become on the same wavelength as the plant. Your energy fields cross and there is a synchronicity between you. Like the Vesica Piscis (see page 32), you merge together and, through the Mandorla, alter your lens of perception.

▲ The communication is energetic, and it arrives first through the light language of sacred geometry. It is then interpreted by your body as a feeling, sensation, memory, idea or vision. It may offer you healing or reveal to you its particular medicine. This may come to you straight away or later on as an 'ah-ha' moment.

▲ The more you repeat this practice, the more sensitive and empathic you become. Intuition is like a muscle that needs to be worked.

▲ You can do this exercise with the same plant at home. As you build up your relationship, you may find yourself watering your plant at seemingly random times or moving its position as the plant 'speaks' to you.

Body

'The only true voyage of discovery…would be not to visit strange lands but to possess other eyes, to behold the universe through the eyes of another, of a hundred others, to behold the hundreds of universes that each of them beholds, that each of them is.'

Marcel Proust (1871–1922), author and critic

Our bodies are a map of the universe, a holographic anatomy that contains the whole of existence within each of its parts. Blood flows through the canyons of our veins, labyrinths trace across our fingerprints, a galaxy twists in the iris of our eye. The proportion of water to land mass on Earth is similar to our own. The electromagnetic field of the Earth casts the same shape as that of our heart. The roots of a tree spiral through the soil like the umbilical cord that attached us to our mothers at birth. Coral dances in the water like the cilia in our lungs. From our conception, we share the same sequence of creation with all organic matter; our organs birth themselves from an inherited blueprint, the codes of our ancestors tightly wound into our sacred geometry, and when we let out our first cry we imprint its signature upon the air.

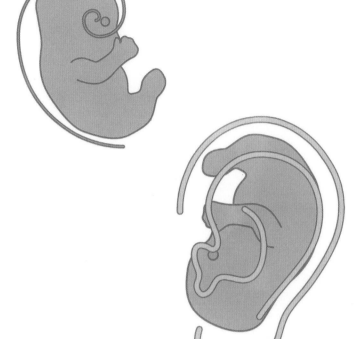

The holographic anatomy of the human body contained within the ear. The head equals the ear lobe.

DNA MIRROR

The spinning helix formation of weather patterns is mirrored in our DNA. When we look at a vertical cross-section of DNA, we see stacked and rotating double pentagons. There are ten rungs in a complete rotation, which echoes the Pythagorean tetractys (*see* page 71) and the ten sephiroths (nodes) of the Kabbalistic Tree of Life (*see* page 43).

In Ayurvedic, Tibetan and Traditional Chinese Medicine (TCM), practitioners work with the flow of energy along meridian lines in the body (like the Earth's energy grid) to release blockages and stagnation, because energetic imbalances create disease patterns. Massage, acupressure, acupuncture and herbs are all part of this system of treatment. Microsystems acupuncture, reflexology and iridology all work on the principle that the whole exists in each part.

Golden proportions of the body

Leonardo Da Vinci drew on the work of Roman architect Vitruvius to map the architecture inherent in the design of the human body and to reveal its golden proportions.

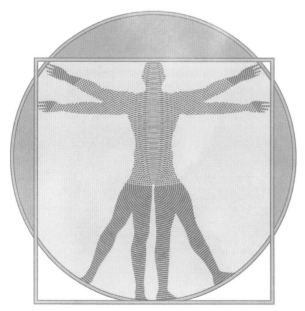

Vitruvian man: the proportions of the human body by Leonardo Da Vinci, influenced in part by the work of Roman architect Vitruvius.

SACRED GYM: GOLDEN PROPORTIONS OF THE BODY

▲ Clench your fist and notice how it forms a spiral. Hold out your hand in front of you and extend your index finger. The Golden Ratio can be found between the knuckle on your hand, the base of your finger and the tip of your finger.

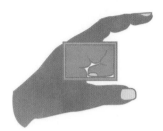

▲ Stretch out your arm. Observe the Golden Ratio in the relationship between your shoulder, elbow and wrist.

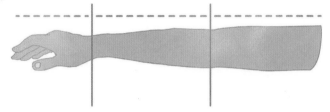

▲ Look in the mirror. Where your nose is in relation to your chin and forehead follows the same approximate Golden Ratio.

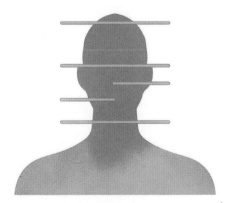

THE PINE CONE IN THE BRAIN

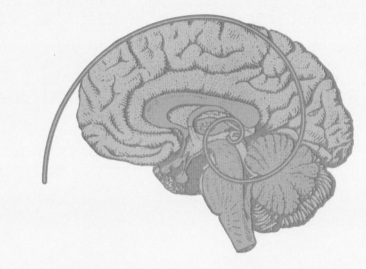

The pineal gland (so-called as it is shaped like a pine cone) is responsible for producing melatonin and taking care of our circadian rhythms, which synchronize our biological cycles with our environment. The gland is connected to our third-eye chakra, found on the forehead between the eyes. The image of a pine cone is found throughout ancient monuments and works of art as representative of spiritual knowledge.

The anatomy of consciousness

Consciousness exists in every cell, every atom of our bodies. We have developed, or rather unravelled, in such a way that we choose to host our consciousness in our intellectual mind. We allow ourselves to be led by our thinking, not by our intuition and instinct, which come from the heart and our subtle fields of energetic perception. If we can recalibrate our inner alchemy to perceive consciousness in other areas of our bodies, to allow our intuition and instinct to process external information before it is transferred to our minds for analysis, we will have shifted our experience to one that is attuned to its environment, the wisdom of nature and the light of the self. Our whole body is an antenna, tuning in to the web of electromagnetic information.

We notice this domino effect of energy when we *feel* another energetic field come into our own and recognize how it has a direct impact on us. This is why brushing past an angry stranger can put us in a bad mood and going for a nature walk can make that same mood melt away. Our environment is in a constant state of energy transfer, stimulus and response, the exchange and transformation of one thing to another. Resonance occurs when two objects vibrate at the same frequency. Everywhere we go we are either in a state of coherence with our environment or we are not, which is known as dissonance.

Mythologist Joseph Campbell (1904–1987) talked of consciousness as the light in a room that is lit by many bulbs. Each bulb or 'vehicle of light' is separate and may vary in quality, but overall one unified light is produced. People, he says, are like light bulbs. We are each 'vehicles of consciousness' and it is the quality of our consciousness that is most important in determining the level of light around us. When our consciousness is not maintained and we are operating at a low vibration, this feeds back into the collective and we see a poor overall quality of consciousness manifesting in our socio-political and economic systems.

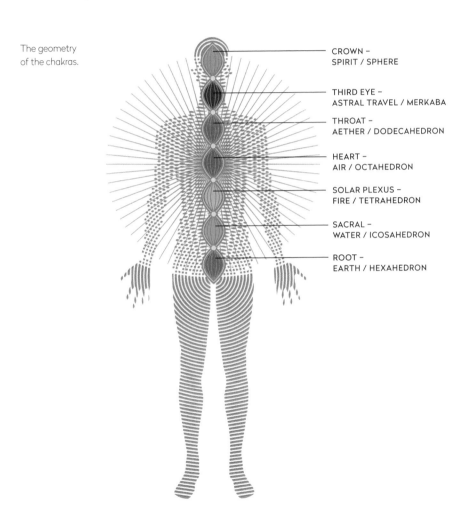

The geometry of the chakras.

CROWN –
SPIRIT / SPHERE

THIRD EYE –
ASTRAL TRAVEL / MERKABA

THROAT –
AETHER / DODECAHEDRON

HEART –
AIR / OCTAHEDRON

SOLAR PLEXUS –
FIRE / TETRAHEDRON

SACRAL –
WATER / ICOSAHEDRON

ROOT –
EARTH / HEXAHEDRON

HEART CENTRE

The heart is the largest generator of electromagnetism in the human body, around a hundred times stronger than the brain, radiating several feet from the body. In a coherent state, the heart is like an amplifier and transmitter, a fine-tuned instrument of electromagnetism capable of receiving and sending messages to the wider field.

Heart torus

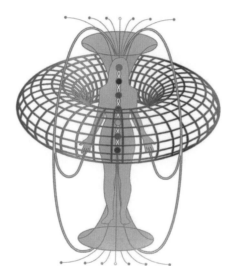

The torus produced by the electromagnetic field of the human heart.

VISUALIZATION: YOUR HEART'S SIGNAL

The seven muscles of the heart form a torus, which makes it our primary sensor when we are communicating through sacred geometry. The way in which your toroidal energy field is spinning directs your overall experience. Tune in to the intelligence of your heart. Enter into it as if it is a sacred cave. Feel the quality of the information that it is giving out and receiving. What is its signal?

The Platonic world year

Occultist and visionary Rudolf Steiner described how the macrocosm of the cosmos is reflected in the microcosm of our body. He described a Platonic year – the time it takes for all the planets and fixed stars to complete a cycle and return to their starting points – as 25,920 years. We are completely in sync with the rhythm of the Earth, which is the source of our biological rhythms. With each breath we mimic the cycles of the cosmos; each day we take a grand cosmic breath.

'On average, a human being breathes 18 times a minute. This may, of course, vary, for our breathing is different in our youth, and in old age, but if we take an average, we obtain as a normal figure for the respiration, 18 breaths a minute. We thus renew our life rhythmically 18 times a minute. Let us see how often we do this in one day. In one hour, this would be equal to $18 \times 60 = 1080$ (*see* page 82). In 24 hours: $1080 \times 24 = 25,920$ times.'

Rudolf Steiner (1861–1925)

COSMIC BREATH

In Vedic science, the lifespan of a universe is one breath of the god Vishnu. With each exhale, multiple universes are birthed into existence, each with its own creator god, Brahma. When the out breath and period of expansion ceases, Vishnu inhales the universes and Brahmas back into the cosmic void, and another breath is exhaled. According to the Dogon tribe of Mali, this inhalation will be experienced as the universe being sucked towards and then collapsing into a supermassive black hole, such as the one at the centre of our galaxy the Milky Way.

Earth magic

'I am the flame above the beauty in the fields; I shine in the waters;
I burn in the sun, the moon, and the stars. And with the airy wind,
I quicken all things vitally by an unseen, all-sustaining life.'

Hildegard von Bingen (1098–1179), Benedictine abbess, writer and philosopher

The ancients understood the sacred geometry of their lands.
They knew the safe passages and the ones to avoid, caves for
redemption and mountains for sacrifice. They knew how to
harness natural energies and at potent points they buried
their dead, made offerings to their gods and erected
astrological monoliths, such as Stonehenge.

Song lines, portals and labyrinths

SONG LINES

Aboriginal peoples use song lines to guide them across Australia. During a ceremonial migration, star songs are sung to navigate the same paths that their ancestors have taken for thousands of years. In their creation myth, also known as the Dreaming, earthbound creator beings formulated the landscape, and song lines follow the same sacred paths of creation. They believe that all information is contained in Dreamtime and that all beings can access this space.

PORTALS

In many mythologies, supernatural beings or animals act as guardians, each one with its own domain, protecting gateways and crossings that bridge dimensions. These portals are said to exist in caves, gorges, springs, waterfalls, at the ends of rainbows. In Iceland, there is an enduring belief in *huldufólk*, hidden elves that resemble people and are capable of opening portals and travelling between realms. In Celtic mythology, fairie folk are enchanters, able to open up parallel worlds through rings, doors and steps to the otherworld.

A portal can be viewed as a physical passage from one place to the next or a place of spiritual transition. On a psychological

---

ELEMENTALS

In paganism and folklore, the beings that guard the elements include:

▲ Air: sylphs and giants
▲ Fire: salamanders
▲ Water: undines and mermaids
▲ Earth: gnomes and fairies

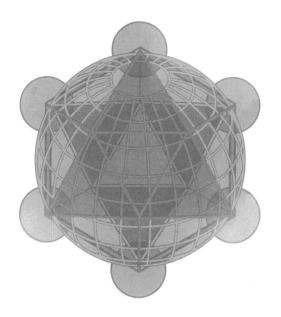

Metatron's Cube and the sacred geometry of the Earth.

level, it can refer to a way of accessing hidden or unknown depths within. Working with sacred geometry is about revealing these hidden passageways that are woven into our fabric.

LABYRINTHS

A labyrinth is a type of portal, symbolic of our life process and the death and rebirth cycle. Labyrinths have been used as passageways for the souls of the dead for many years. The labyrinths I visited on one of the Solovki Islands, an archipelago in the White Sea, Northern Russia, were made with grids of stones. Indigenous people from the mainland would travel to the island to bury their dead, believing the place to be where this world ended and the next began. The labyrinths would honour the spirits of the land and open doors for their beloved ones.

CREATE A LABYRINTH

A labyrinth created with a single line (unicursal) has just one path in and out and its simplicity allows for its navigation to be a meditative process. It is traditionally made of seven rings held within the eighth boundary, reminiscent of human embryology and the chakras.

▲ First draw your template. You can do this using a pen or a medium like sand or soil.

▲ If you like, you can scale it up and create it outside, working with the natural lay of the land and landmarks, such as trees or rocks.

After time, many labyrinths fade away, the stones erode or are moved by unknowing hands, but the energetic imprint remains on the land. This is the case with the mystical labyrinth at Glastonbury Tor, where initiates still trace its path. Next to the Hawara pyramid, south of Crocodilopolis, in Egypt, is the site of what is considered to be the original labyrinth, the prototype of the mythical labyrinth of Crete built to house the Minotaur. Nothing remains of it now, but it is possible to trace its energetic pathways, as I did, and tune in to its complex resonance.

Labyrinths, by design, are symbolic of our intestines. In turn, they reflect the ways in which we digest our experiences, offering an invitation to let go and release all that is no longer serving us and opening us to receive deeper spiritual nourishment and move forward on an aligned path.

HOW TO WALK THE LABYRINTH

▲ Take a moment before entering to call upon your shadow self, outdated emotional patterns or trauma responses that no longer serve you.

▲ Face them as you walk into the labyrinth. You do not need to dwell on them, it is just the vibrational pattern that you are tuning in to.

▲ Offer them up to the labyrinth for transformation and change.

▲ Find your still centre when you reach the middle of the labyrinth. This is the moment of death and rebirth.

▲ Give thanks to the energetics of the space for the opportunity for healing and release.

▲ Tune in to the labyrinth, the sacred geometric form you have created, allowing it to energize and lift you as you turn back and walk into the light of a new field of awareness.

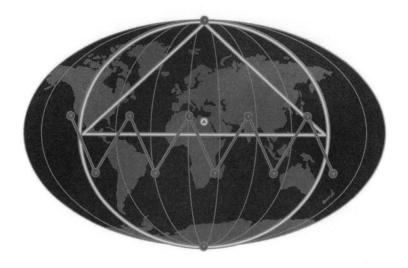

Vile vortices (left to right): Hamakulia volcano, east of Hawaii; Easter Island megaliths; Bermuda Triangle; South Atlantic anomaly, east of Rio de Janeiro; Algerian megaliths; Zimbabwean megaliths; Indus Valley, Pakistan; Wharton Basin; Devil's Sea, south of Japan; Loyalty Islands, New Caledonia; North and South Poles.

Vortices

These occur where the flow of air, water or energy spins around an axis that exerts a strong downward force on anything that comes into contact with it. Many ancient sites can be found in these places.

The unexplained disappearance of boats and planes has occurred at 12 documented sites, which are also known as 'vile vortices' (a phrase coined by biologist and writer, Ivan T Sanderson), along the Tropics of Cancer and Capricorn and at the North and South poles. The most famous site is the Bermuda Triangle. These unfortunate events are reportedly caused by an increase in the Earth's electromagnetic field at these junctions. Connecting up the 12 vortices creates a three-dimensional icosahedron over the Earth (*see* page 121). The icosahedron is associated with water and the majority of the 12 sites lie over the sea or snow-capped mountains.

TELLURIC ENERGY

Electrical currents that run along curved pathways either underground or in the sea produce what is known as telluric energy. Excessive build-ups of energy – gas or water – are released through geysers, volcanoes and earthquakes. This system is reflected in our own bodies, using the pressure points and meridian lines of Eastern medicine.

Power spots

Caves, springs, whirlpools, waterfalls, mountains and volcanoes may all harbour power spots. These occur where two areas of land, or water, with different conductivity rates meet, producing a concentrated electrical charge. The double spiral occurs in nature when two different currents of electricity or water meet, especially when the water is high in minerals or metals, such as iron ore.

'The real Earth, viewed from above, resembles a ball made of twelve pieces of leather, variegated and marked out in different colours.'

Plato (*c.*427–347 BC)

Plato appears to be suggesting that the 'real Earth', meaning its energetic structure, is a dodecahedron (12-sided shape), with all of the solids nested within it. Visionary quantum theorist Nassim Haramein (b.1962) supports this concept, and it is also in sacred geometer John Michell's (1933–2009) work on the ancient creation myths of the original 12 tribes of the Earth. It is clear from the positioning of many sacred sites that the ancients were aware of the body of the Earth as an organism, and placed their temples on these power spots. It is no coincidence that many of these sites link up, indicating that there was a worldwide exchange of knowledge.

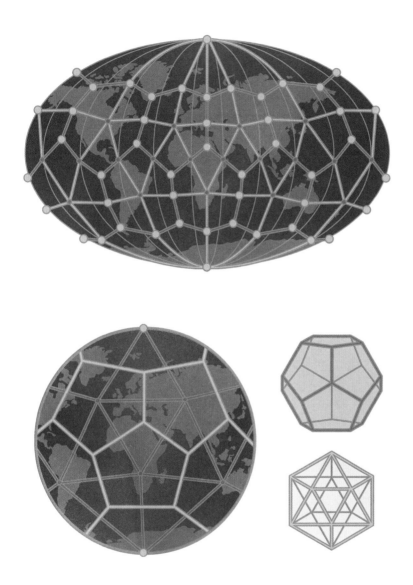

Dodecahedral and icosahedral grid systems based on Crystal Earth Theory (1975) of Nikolai Goncharov, Vyacheslav Morozov and Valery Makarov.

Evidence of Earth's energies

▲ Circles of standing stones exist all across the globe. In Britain, prominent examples are Avebury and Stonehenge in Wiltshire and Callanish on the Isle of Lewis, Scotland. How and why they are there is still open to debate but none arrived by accident. Each one is astrologically aligned, particularly to the Sun and Moon; the solstices and equinoxes are the most consistent points of reference. Attuned to the skies with phenomenal precision, these stones are also aligned with telluric currents and waterways that create vortices of energy. It is in this way that they function as amplifiers, instruments of vibration and transmitters of earthly and cosmic forces.

▲ Large-scale, often complex, geometric patterns appear in crop fields in and around specific sites, such as Avebury and Stonehenge. Often these are accompanied by UFO sightings and have for a long time been a mysterious fascination. Enthusiasts believe that the crops' energetic signature is altered outside of space-time through a change in wave pattern. Crop analysis of key circles has shown that the plant is not damaged and continues to grow in its bent form, implying that the change occurs on a cellular level.

▲ The Nazca Lines of Peru are large-scale geoglyphs etched into the land. They are vast geometric shapes; some lines run for over 48 kilometres (up to 30 miles). Their significance is still being debated, but again we see a precise astronomical alignment. Professor Paul Kosok (1896–1959) called the site the 'largest astronomy book in the world'.

▲ In 1921, Alfred Watkins (1855–1935) realized that many sacred sites, geographic points of interest and power spots in Britain could be linked with straight lines, which he called ley lines. These came to be regarded as the equivalent of the fairie passes of European folklore. Many phenomena have been observed around ley lines and particularly at their intersections (Stonehenge lies exactly at one such intersection). These include spikes in the activity of birds, insects and weather and the appearance of those mysterious crop circles.

▲ In China, a similar concept exists in *lung-mei* or dragon lines, believed to be the meridian lines of the Earth and the paths of the dead. For the same reason, man's creation of symmetry and straight lines is regarded with suspicion in Japan. Chinese geomancy (meaning to 'divine the Earth') is called feng shui, which translates as wind-water. It is concerned with the energetics of space and the dragon lines that allow spirits and unseen forces to travel. Feng shui is built on the belief that these energies must be allowed free passage, otherwise places can become corrupted by negative spirits or stale energy.

Earth's toroidal field

When the Earth spins between its two magnetic poles, it creates a field of electromagnetism. This produces a toroidal field, as it creates the sacred geometric shape known as the torus, which looks like a donut. The torus is a primary energy pattern where energy flows in one end and out the other, then folds back in on itself. Plants, animals, humans, planets and galaxies, even our blood cells – every living thing – has an energy field matrix or torus, which communicates with its immediate environment and feeds back its findings to the central system.

Research into this has been pioneered by the Institute of HeartMath and its Global Coherence Project, which suggests that the Earth's magnetic field is a living library or medium for bioelectromagnetic information transfer between all systems of life within it.

Opposite: Torus tube created by the Earth's toroidal field.

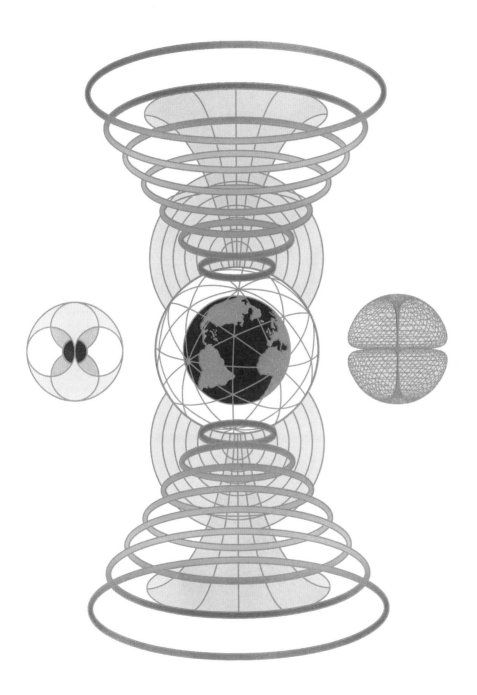

Geopathic stress

Systems of geomancy are concerned with the impact on
the body of prolonged periods of exposure to power spots,
such as sleeping or living on cross-points of Earth grids.
These are thought to change the voltage of the body,
causing psychological and physiological stress.

When the Schumann resonance of the Earth is disturbed,
through the movement of water, mineral deposits, fault lines
and other geological phenomena, the frequency of 7.83Hz is
distorted. Geomagnetic storms can drastically alter this wave
and research has shown that this has a particular impact on
the heart, altering blood pressure, heart rate and melatonin
levels in the brain. Spikes in suicide rates and cardiac arrest
during periods of solar flares have been documented.

We are electromagnetic beings, using light and movement to
communicate information through our own internal systems
and biochemistry. Yet we are living in an electromagnetic
construction of our own making to facilitate external
communication that is in conflict with the natural order, and
we are having to adapt to this. Nature communicates through
the electromagnetic spectrum, which is becoming distorted.
Bees and birds that use electromagnetism for navigation are
losing their way, their internal compass scrambled. We need
to pay close attention to our own barometers and be sure to
cleanse our nervous systems regularly with time in nature.

Ceremonies and practices to appease this imbalance and
honour the land include offering blessings to the elemental
beings, laying crystal grids in sacred geometric forms,

'earth acupuncture' using needles carved from yew or other magical woods and applying the principles of Chinese feng shui to compensate for energetic imbalances.

Crystal grids

Laying crystals in a sacred geometric form creates a hyper-dimensional energetic structure that holds and amplifies its vibration to create an electromagnetic field of potential. Any of the shapes we have discussed in this book are suitable to make a grid and there is no need to be constrained to preordained forms. Intuitive freestyle grids can be just as effective.

LAY A CRYSTAL GRID

▲ When you have chosen your grid and crystals, find an appropriate space where they will be undisturbed.

▲ Ensure that you are using crystals that have been energetically cleansed. You can do this with water, placing them in moonlight or sunlight, through energy healing or by smudging.

▲ Clarify your intention for the grid.

▲ Consciously lay the crystals. If you are working with a spiral, for example, choose to work from the inside out to release stagnant energies, or from the outside in to call in new energies and flow. If you are working with a grid, draw the grid and place the crystals at the intersections.

▲ Activate the grid using a crystal wand, your hand, or your mind's eye to join the crystals together into the shape of the grid.

▲ When the grid has served its purpose, take it apart in the reverse order from which it was laid.

▲ Cleanse the crystals and the space using any method mentioned above.

Design

'Architecture is crystallized music.'

Johann Wolfgang von Goethe (1749–1832), scientist, writer, critic and diplomat

Architecture is the living expression of the consciousness of the people of its time. The ancients built places of worship, community and healing in response to the pure energetics and lay of the land. These varied from simple round temples mirroring celestial bodies and standing stones employing precise astronomical alignment, to vast complexes of supremely intelligent design that served as power houses of consciousness. Devotional spaces were designed to channel higher vibrations to bring worshippers closer to spirit and enter into communion with their gods.

TIME AND SPACE

Mythologist and symbologist Robert Lawlor draws a connection between 'temple', 'tempo', 'tempest' and 'template' and the notion of time and space. The origins of the temple lie in providing a space for oracles and prophecy. When we look at the way in which many ancient temples were built, we can see that they were designed using sacred geometry to create complex astronomical clocks for predicting cycles and so 'seeing into the future'. Time and space were considered to be active forces, not just measuring devices, and honoured as such.

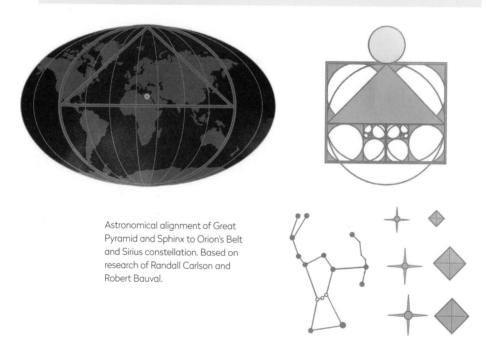

Astronomical alignment of Great Pyramid and Sphinx to Orion's Belt and Sirius constellation. Based on research of Randall Carlson and Robert Bauval.

Sacred geometric principles were used to amplify existing energies for prayer and contemplation, protection and sanctuary. Architects were concerned with maintaining a balance between attracting certain energies and containing these within their boundaries and creating barriers to protect from less fortuitous ones.

The Ancient Egyptians began their calendar with the heliacal rising of the star Sirius, which heralded the annual flooding of the Nile. This occurred each year when it became visible above the horizon, just before dawn, after a period of absence. Rising just after the constellation of Orion – associated with Osiris, god of the underworld – Sirius represented the goddess Isis and was a symbol of rebirth.

Using the Earth's energetics

Sacred sites were aligned to the cosmos as astronomical timekeepers, geomagnetic resonators attuned to the celestial bodies. Stonehenge has 29 stones of equal width and one of half width, representing the 29.5 days between full moons, and it is aligned to the sunrise of the summer solstice and the sunset of the winter solstice. Many other sites track the equinoxes and solstices, including Newgrange in Ireland, Machu Picchu in Peru and Chichen Itza in Mexico. Many of these sites follow the same proportions as each other, including the Golden Ratio (*see* page 75), and track the same poles and Earth grids, implying that these ancient civilizations had a common purpose and universal understanding of the energetics of the Earth.

Relationships of ancient sites based on research of Hugh Newman and J Martineau's World Grid Program.

Celestial correspondence to the land was just one aspect of working with sacred geometry. The methods of construction – stacking, packing, interlocking, gridding, spiralling – have often been inspired by natural order.

▲ Rudolf Steiner was moved to create 'Goetheanum' buildings in honour of Goethe's work with organic structure and metamorphosis.

▲ Antonio Gaudi observed natural forms and took inspiration from them for his works, including the Sagrada Familia in Barcelona.

▲ Zaha Hadid described her Aquatics Centre for the 2012 Summer Olympics in London as 'inspired by the fluid geometry of water in movement'.

Measurement

Temples are built by humans for humans, so it follows that the human form also provided a template for architecture and the spiritualization of matter. To Michelangelo, knowledge of anatomy was vital to architecture. The temple creates the energetics for the physical manifestation of non-physical entities, such as gods and goddesses, through iconography, statues and architecture. Goddess temples in Malta were shaped as voluptuous breasts and full bellies. In his book *The Temple in Man*, RA Schwaller de Lubicz (1887–1961) maps out the human body, including organs and glands, and notes how its proportions contain the measurements of the whole universe.

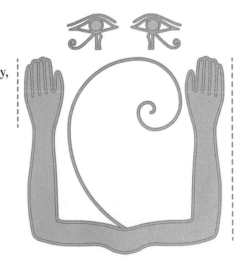

Measurement itself was a magical process and only certain initiates, priests and priestesses were able to carry out the task. The approval and instruction of higher powers were called in to assist with the work, astrological calendars were drawn, offerings were made to the gods and long periods were spent in prayer and meditation in preparation. In Ancient Egyptian mythology, the goddess Seshat rules over geometry, architecture and writing as a counterpart to Thoth. She would create temple layouts by stretching a sacred chord (*see* image, page 28). Builders used knotted ropes to scale and measure the correct proportions and they would call on Seshat for guidance.

Egyptian measuring systems (opposite); pyramid proportions (below); the temple in man (right).

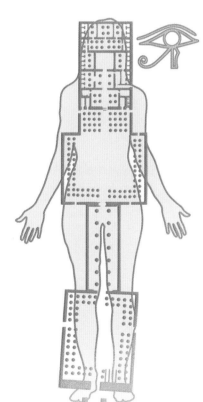

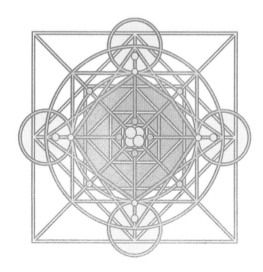

Other geometric principles prevalent in sacred architecture are the Golden Ratio (*see* page 75), spiral (*see* page 90), triangle, double square, circle within the square, and the square roots of one to five – all of which we can find in the Great Pyramid of Giza.

UNITS OF MEASUREMENT

Units of measurement were and still are in some cases based on the human body, reflecting the intuitive understanding of the correspondence between the microcosm and macrocosm. Feet and inches are based on a human foot and thumb. The Ancient Egyptians measured in whole numbers which were sacred in themselves. Whereas when we measure using the modern decimal system, we get fractions.

▲ Digit – width of finger
▲ Palm – four digits/ crown of skull
▲ Cubit – extension from hand to elbow
▲ Royal cubit – one-sixth longer than a cubit

Square roots

The square roots of 2, 3 and 5 are found in the Vesica Piscis. Square roots are fundamental to creation and were employed by the ancients in constructing buildings.

In his book *Sacred Geometry, Philosophy & Practice,* Robert Lawlor looks at the spiritual and psychological motivations driving square roots. The root of a square is hidden beneath the surface, just like the root of a tree. It nourishes the shape and its emergence as a root is deeply transformational; from the depths it breaks down matter in order to release nutrients and provides nourishment and strength to the life above. In this way, square roots provide signposts to deeper truth and wisdom contained within.

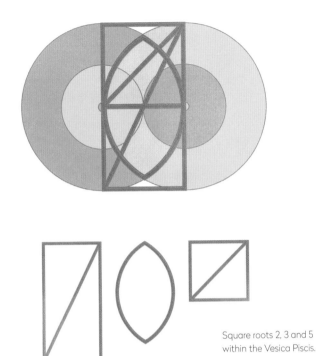

Square roots 2, 3 and 5 within the Vesica Piscis.

ROOTS

▲ Square root of 2. If we divide the body so that it follows the square root of 2 (1.41421356237), we come to what in Eastern philosophy is known as the *dantian* or *hara* – just below our belly button, our sacred axis, our central point, considered to be the place from where our vital force emanates.

▲ Square root of 3 is found in the Mandorla of the Vesica Piscis

(see page 32). A lens into the birthplace of creation and the divine feminine.

▲ Square root of 5 is embedded in *phi* and growth patterns in nature. It is found in the double square and used in sacred architecture, including the King's Chamber inside the Great Pyramid at Giza.

SQUARING THE CIRCLE

This is the holy grail for sacred geometers. Jung called it 'the archetype of wholeness'. To square the circle means to create a square with a perimeter that is identical to the circumference of the circle, which is technically impossible because the circumference of the circle will always be an irrational number. Symbolically, squaring the circle is to unite spirit with matter – to spiritualize matter.

Interpretation

The more that we investigate the sacred geometry encoded in sacred places, the more multi-layered they reveal themselves to be. In esoteric lore, the secrets of the universe are concealed within these buildings and the task of translation was given to skilled initiates, such as the Knights Templar, Rosicrucians and Freemasons. In order to truly understand sacred geometry, we need to read between the lines and join up the dots. It is not about the seen but the unseen. In this way, buildings can be *read* like oracles.

The task of the architects of sacred space was to create the right harmonics for prayer but also for healing. Asclepius, Greek god of healing and medicine, created temple spaces designed for healing and transformation. Pilgrims would enter into a period of fasting and meditation before lying down to sleep on temple altars, drifting into a receptive state of consciousness, in communion with the energetics of the space and resulting in visualizations of medicinal plants and guidance through dreaming.

An initial pilgrimage is often required to test the pilgrim's quality of spirit. Then the journey of the soul is played out in the temple's architecture.

- ▲ At the Zentsuji Temple in Japan, a 100-metre (328-feet) pitch-black passageway leads to enlightenment.
- ▲ Passageways and tunnels connect the rock-cut churches of Lalibela, Ethiopia.
- ▲ Gnomonic spirals are a feature of the step pyramids of Mesoamerica and Egypt.
- ▲ The counter-clockwise turn around the Rabba at Mecca.
- ▲ The turning of prayer bells around the parameters of Hindu and Buddhist temples.
- ▲ Pilgrims walk the labyrinth of Chartres Cathedral in France.

- ▲ The Borobudur Buddhist Temple complex in Indonesia uses 3, 6 and 9, sacred and abundant numbers, revered through the ages. Nine stacked platforms, three circular and six square, ascend to the central dome, which itself has 72 statues of the Buddha. Pilgrims must walk around the base in a clockwise motion, moving upwards through the three realms (Tridhātu) of Buddhist cosmology – the Ārūpyadhātu, the Rūpadhātu and the Kāmadhātu. All of these are designed for contemplation and purification before entering into the sanctity of the holy inner chambers and the presence of idols, relics and sacred objects.

'If you only knew the magnificence of the 3, 6 and 9, then you would have the key to the universe.'

Nikola Tesla (1856–1943), engineer and inventor

SACRED SYMBOLISM

Sacred sites are adorned with religious and spiritual symbolism.

▲ At Tarxien temples in Malta, we see the double-spiral etched into the rock and in the Ħal Saflieni Hypogeum (c.3200 BC) a series of spirals and hexagons painted in oxblood.

▲ The Apprentice Pillar at Rosslyn Chapel, Scotland, appears to depict the Tree of Life and the double helix.

▲ The rock-hewn churches of Lalibela, built in the 12th century by the king as a new Jerusalem, reveal a crossroads of faith with the Vesica Piscis, Fruit of Life, Orthodox cross, Seal of Solomon, the Maltese cross (linked to the Knights Templar) and the Hindu swastika, symbolizing motion.

Symbolic art

When we observe a work of art, what we see with our eyes and recognize with our minds is only a fraction of what is happening on an energetic and biochemical level. The visual cortex is stimulated, neurons fire in the brain and oxytocin rushes through our system as we experience appreciative joy. Beyond that, our unconscious mind is processing the deeper, multi-faceted nature of artistic expression. In a similar way to the principle of vibrational medicine, the body is responding to the patterns and energetics woven into the artwork.

It is part of the human condition to express oneself through art. Creation is a way of connecting with what it is to be human and the quest for meaning, and also of communing with the sacred self within. Stories have been etched into rocks, painted on bodies, carved into structure, and religious iconography created to manifest the spirit of an idol.

Religious art employs the principles of sacred design to create energetic works that are attuned to a saint, god or icon and the essence of their role. Sacred symbols are often used as instruments of meditation, to hold the quality of a particular energy field and achieve altered states of consciousness. In Islam, Allah must never be depicted, so instead visual prayer is expressed through geometric form to evoke the nature of god as creator.

The Sri Yantra is a symbol that is used in the Hindu faith as a meditation device and energetic transmission for spiritual expansion.

Sri Yantra

There are many types of yantra, or sacred diagram, but the Sri Yantra is thought of as the mother of all. It has been around for 12,000 years and is referred to as the holy instrument or instrument of creation. In Sanskrit, the name translates as 'contraption' or 'machine'. The Sri Yantra is aligned with the divine proportion and contains within it the Merkaba and the Flower of Life. It is made up of nine interlocking triangles, four upward pointing signifying Shiva, the masculine principle, and five downward pointing signifying Shakti, the female principle. In its three-dimensional form it is thought to be the mystical and metaphysical Mount Meru or Holy Mountain at the centre of the universe. The Sri Yantra is a powerful meditation tool and contemplative expression of the mechanics of the universe.

SRI YANTRA MEDITATION

In 1987, scientist Alexy Pavlovic Kulaichev (b.1945) of Moscow University measured the brain-wave patterns of people meditating on the Sri Yantra and found that it induced meditative states within the alpha-theta range.

The central point, or *bindu*, of the Sri Yantra is where to focus your attention during meditation. The drawing is an invitation for the observer to witness their own centre. This focusing is thought to activate and decalcify the pineal gland and balance the left and right hemispheres of the brain to deepen meditation and insight. At the same time, it can release the tension and negative charge of old emotional patterns so that you can clear the pathway for manifestation.

Mandalas

Like yantras, mandalas are associated with cosmic consciousness. They originated in the Buddhist and Hindu faiths and usually represent the universe through complex – often geometric – visual designs. Early monks and yogis would draw circles around themselves when they sat outside to meditate, to create a sacred space of alignment, a living mandala. Many use the practice of creating mandalas, often with grains of coloured sand, as a form of meditative prayer.

Psychoanalyst Carl Jung used mandalas in his practice with his patients, seeing them as a mirror to the soul, a 'living conception of the self'.

'Our psyche is set up in accord with the structure of the universe and what happens in the macrocosm likewise happens in the infinitesimal and most subjective reaches of the psyche.'

Carl Jung (1875–1961)

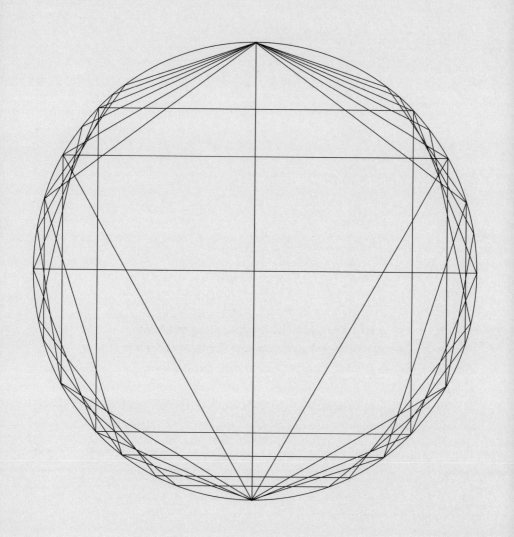

Meditative geometry

'It is through geometry that one purifies the eye of the soul.'

Plato (*c*.427–347 BC), philosopher

The act of drawing sacred geometrical shapes should be a meditative process. Observe yourself and what comes up for you. I will take you on this reflective journey as we go. Using a compass and a straight edge rather than measurements allows for the organic unfolding of one shape from another, as is the natural process of things.

You will need: a compass, ruler or straight edge, pencil and paper. You may choose to use graph paper or a protractor as a guide to accurate angles and straight lines.

1. Draw a circle with your compass and pencil. The compass point becomes your central axis X. Maintain your compass width.

2. Divide the circle into two semicircles by drawing a vertical line through X, creating points A and B. You may also create this division by drawing a horizontal line through point X and marking points C and D on the circumference. (You could then draw a second circle by placing your compass point on point C or D to create a Vesica Piscis – *see* page 40 for how to develop further shapes from the Vesica Piscis.) If you are using a protractor, mark up each 90-degree angle to ensure accurate points.

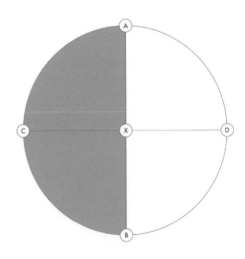

3. To make a triangle, draw a circle with points A and B marked, as in 2. Place the compass point on point B and draw a second circle that intersects the original circle on each side to give points C and D as the feet of your triangle. Now join A, C and D together.

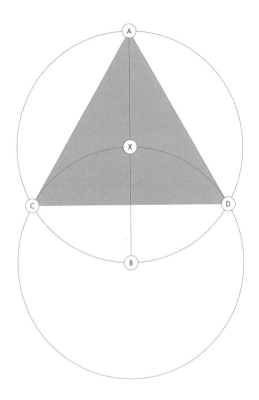

4. To make a square, draw a circle and mark points A, B, C and D, as in 2. Place your compass point at A and draw another circle. Repeat the process at points B, C and D. Each of the four peripheral intersections that you have created provides you with a corner of your square and may now be connected using a straight line.

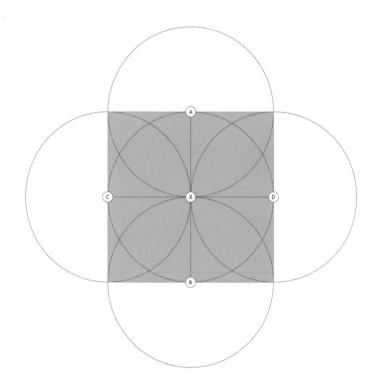

5. To make a pentagon, draw a circle and mark lines A–B and C–D, as in 2. Take your compass and intersect the edges of the circle to give you points E and F from C, and G and H from D. Using a ruler, mark where the lines between E and F and G and H cross the C–D line to give you points I and J. Adjust your compass so that it is the width of the distance from points I to C and J to D. Now draw two circles using I and J as the axis points for your compass. Place your compass on point B and adjust the width to make a circle that just touches the base of each new circle. Where this circle meets the original circle, make a mark for points K and L. Join points K and L together with a line and set your compass to this width. Now mark where the compass touches the original circle from points K and L; this gives you points N and M. Now, to make your pentagon, join up points A-M-K-L-N-A.

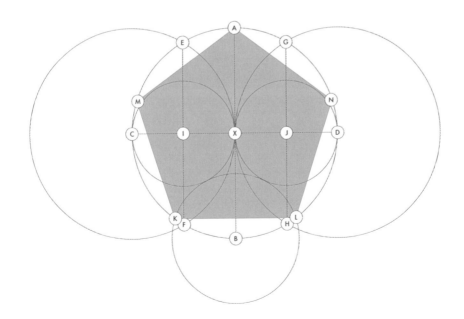

6. To make a hexagon, draw a circle with points A–B marked, as in 2. Move the compass to point A and make a circle forming points C and D, where the new circle crosses the original one. Do the same from point B to create points E and F. Now place your compass at point D to draw a circle that intersects points A and F, then move the compass to point F to draw a circle that intersects points B and D. Continue doing this in a clockwise motion to form a six-petal flower. The tip of each petal is a point of your hexagon. Now join them up.

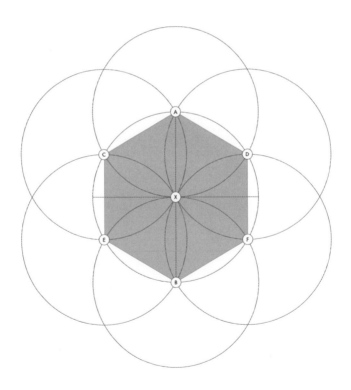

7. To make a heptagon, first make a square from your circle, as in 4. With your A–B and C–D lines marked up, label the four corners of your square clockwise from the top left to get points E, F, G and H. Set your compass to the length of one side of your square and, from points E and F, draw two circles with your compass that intersect the A–B line to create point I. Connect points E and F to I to form points J and K where they cross the original circle. Set your compass to the distance between points A and K; this is now the length of each side of your heptagon. Move your compass clockwise, starting with K, to intersect the original circle at point L, then from L make M, and so on, until you have points N and O. Now join up points A–K–L–M–N–O–J–A.

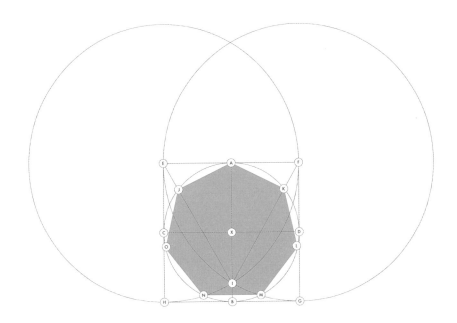

8. To make an octagon, draw a circle with the vertical A–B and horizontal C–D lines marked, as in 2. With your compass on the same setting, draw four more circles from these points as you did when making your square (*see* 4). Now draw two diagonal lines joining opposite corners of the square to create the spokes of a wheel so that your circle is now divided into eighths. Label the points where these lines cross the original circle as E, F, G and H. Now join points A–E–D–F–B–G–C–H–A.

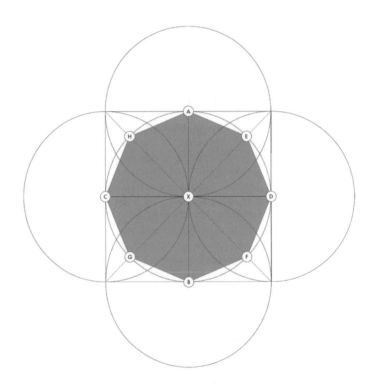

9. To make a nonagon, create a circle with points A, B, C and D marked, as in 2. Set your compass to the distance between A and D then, from point A, draw a circle joining points C and D to create E at the intersection with the A–B line. Place your compass on the line around halfway between points E and B and adjust the width of your compass until it will make a circle that runs through points E and B. From point A, draw a line so that it 'kisses' the edge of this circle to make points F and G. Join points F and G to form the base of your nonagon. Adjust your compass to the width of F–G and mark clockwise around your circle to make points H–I–J–K–L–M. Now join up points A–K–L–M–G–F–H–I–J–A.

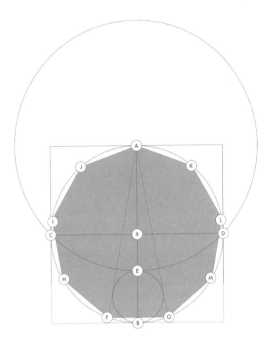

10. To make a decagon, first draw a pentagon (*see* 5). Draw a line from each point through X to the other side to create ten intersections with the outside circle. Connect each of these points to create your ten sides.

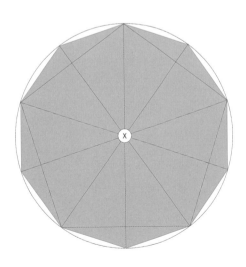

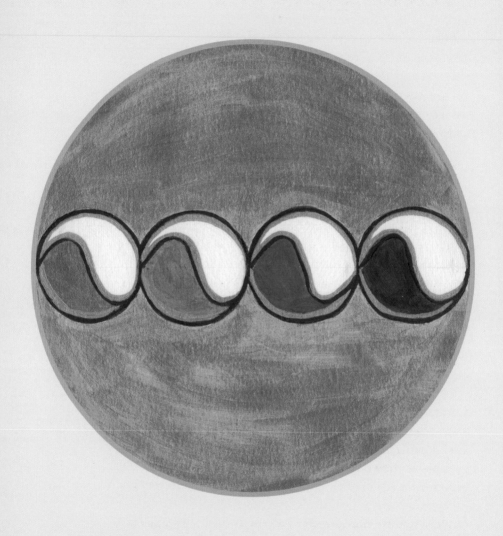

Conclusion

'Will transformation. Oh be inspired for the flame in which a Thing disappears and bursts into something else; the spirit of re-creation which masters this earthly form loves most the pivoting point where you are no longer yourself.'

The Sonnets to Orpheus, Rainer Maria Rilke (1875–1926), poet and novelist

Sacred geometry is a philosophy and a practice. Now that you have mastered its vocabulary, you will start to see patterns unfolding in your daily life and experience more synchronicities. Things are less likely to get lost in translation because you can speak the language. So pay attention, look out for the signs and see what the universe is trying to tell you.

The more you engage with the practice and raise your vibratory field, the more sensitive you will become to energetics. You will be more able to discern which energies are supportive to you and the ones that drain you. You will be more empowered as the architect of your reality and become experienced in shifting your field of awareness.

Explore nature, immerse yourself in its geometry and be in conversation with it. Talk to your plants! It is an osmosis of consciousness. As you move through life and respond to your experience, you reveal your own unique sacred geometry to the world and open yourself up to receiving its wisdom.

References

INTRODUCTION

Hawking, Stephen, *A Brief History of Time*, Bantam Books, 1988, p.174

Jung, Carl G, *Man and His Symbols*, Dell Publishing, 1964

MOVEMENT AND FORM

Bentov, Itzhak, *Stalking the Wild Pendulum: On the Mechanics of Consciousness*, Destiny Books, 1988, p.110

Campbell, Thomas, *My Big TOE*, Lightning Strike Books, 2003

Capra, Fritjof, *The Tao of Physics*, Flamingo, 1982

Lanza, Robert, *Biocentrism*, Benbella, 2009

Talbot, Michael, *The Holographic Universe*, Harper Collins, 1996

Whicher, Olive, *Projective Geometry: Creative Polarities in Time and Space*, Rudolf Steiner Press, 2013

CREATION

De Santillana, Giorgio; von Dechend, Hertha, *Hamlet's Mill: An Essay Investigating the Origins of Human Knowledge and Its Transmission Through Myth*, David R Godine Publisher, 1977, p.60

Fuller, Buckminster, *Critical Path*, St Martin's Press, 1981

Michell, John; Brown, Allan, *How the World is Made: The Story of Creation According to Sacred Geometry*, Thames and Hudson, 2012

Sheldrake, Rupert; McKenna, Terence; Abraham, Ralph, *Chaos, Creativity and Cosmic Consciousness*, Park Street Press, 2001, p.20

NUMBER

Blavatsky, Helena P, *Isis Unveiled Vol I*, Cambridge University Press, 1877, p.290

Bohm, David, *Wholeness and the Implicate Order*, Routledge, 1980

Hall, Manly P, *The Secret Teachings of All Ages*, Tarcher/Penguin, 2003

Heath, Richard, *Sacred Number and the Origins of Civilisation*, Inner Traditions, 2007

Javane, Faith; Bunker, Dusty, *Numerology and The Divine Triangle*, Whitford Press, 1979

Jodorowsky, Alejandro, *The Way of the Tarot*, Destiny Books, 2009

Schneider, Michael S, *A Beginner's Guide to Constructing the Universe*, HarperCollins, 1994

SOUND

Cage, John, *Silence*, Marion Boyars, 2011

Emoto, Masaru, *The Hidden Messages in Water*, Atria Books, 2005

Hazrat Inayat Khan, *The Mysticism of Sound and Music*, Shambhala, 1996 (revised edition)

Jenny, Hans, *Cymatics, Vols 1 and 2*, Macromedia, 1967, 1974

Lawlor, Robert, *Sacred Geometry: Philosophy & Practice*, Thames & Hudson, 1982

NATURE

Buhner, Stephen Harrod, *The Lost Language of Plants*, Chelsea Green Publishing, 2002

Capra, Fritjof; Luisi, Pier Luigi, *The Systems View of Life: A Unifying Vision*, Cambridge University Press, 2014

Edwards, Lawrence, *The Vortex of Life: Nature's Patterns in Space and Time*, Floris Books, 2018

Gordon-Lesmoir, Nigel, *The Colours of Infinity: The Beauty and Power of Fractals*, Springer, 2010

Harding, Stephan, *Animate Earth: Science, Intuition and Gaia*, Green Books, 2013

Heath, Richard, *Sacred Geometry in the Realm of the Planets*, Inner Traditions, 2002

Livio, Mario, *The Golden Ratio: The Story of Phi, the Extraordinary Number of Nature, Art and Beauty*, Headline, 2002

Mandlebrot, Benoit, *The Fractal Geometry of Nature*, W H Freeman and Company, 1977

Narby, Jeremy, *The Cosmic Serpent: DNA and the Origins of Knowledge*, Phoenix, 1995

Schmitz, Eckhart R, *The Great Pyramid of Giza: Decoding the Measure of a Monument*, Roland Publishing, 2012

Steiner, Dr Rudolf, 'Man's position in the cosmic whole', lecture delivered in Dornach (Switzerland), 28 January 1917

Warm, Hartmut, *Signature of the Celestial Spheres: Discovering Order in the Solar System*, Sophia Books, 2010

BODY

Childre, Doc; McCraty PhD, Rollin, *The Appreciative Heart: The Psychophysiology of Positive Emotions and Optimal Functioning*, Institute of HeartMath, 2002

Lipton, Bruce, *The Biology of Belief*, Hay House, 2005

EARTH MAGIC

Goncharov, Nikolai; Morozov, Vyacheslav; Makarov, Valery, 'Is the Earth a Large Crystal?', *Khimiya i Zhizn* (*Chemistry and Life*), USSR Academy of Sciences, 1973

Michell, John, *The Dimensions of Paradise: Sacred Geometry, Ancient Science and the Heavenly Order on Earth*, Inner Traditions,1972

Michell, John, *The New View Over Atlantis*, Thames & Hudson, 1983

Rhone, Christine; Michell, John, *Twelve-Tribe Nations: Sacred Number and the Golden Age*, Inner Traditions, 1991

Sanderson, Ivan T, 'The Twelve Devil's Graveyards Around the World', *Saga* magazine, 1972

Three Initiates, *The Kybalion – Centenary Edition: Hermetic Philosophy*, TarcherPerigee, 2018

Vidler, Mark; Young, Catherine, *Sacred Geometry of the Earth: The Ancient Matrix of Monuments and Mountains*, Inner Traditions, 2016

DESIGN

Althaus, Karin; Muhling, Matthias; Schneider, Sebastian (eds), *World Receivers* (exhibition catalogue), Hirmer Publishers for the Lenbachhaus Museum, 2018

Bauval, Robert; Gilbert, Adrian, *The Orion Mystery: Unlocking the Secrets of the Pyramids*, Three Rivers Press, 1994

Cirlot, J E, *A Dictionary of Symbols*, Welcome Rain, 1971

Jung, Carl G, *Mandala Symbolism*, Bollingen Series, Princeton University Press, 1972

Khanna, Madhu, *Yantra, The Tantric Symbol of Cosmic Unity*, Thames & Hudson, 1979

Lawlor, Robert, *Sacred Geometry: Philosophy & Practice*, Thames & Hudson, 1982

Lawlor, Robert, *The Geometry of the End of Time*, Robert Lawlor, 2015

Mann, A T, *Sacred Architecture*, Element Books, 1993

Newman, Hugh, *Earth Grids: The Secret Patterns of Gaia's Sacred Sites*, Wooden Books, 2008

Schwaller de Lubicz, R A, *The Temple in Man: Sacred Architecture and the Perfect Man*, 1981

CONCLUSION

Rilke, Rainer Maria, *The Sonnets to Orpheus*, Vintage Books, 2009, XII, p.157

Index

Acknowledgements

Teachers and mentors:
Fritjof Capra, Bonny Casel, Patricia Lehman Awyan,
Bruce Lipton, A T Mann and Dr Yubraj Sharma.

Publishing and proofing:
Valeria Huerta, Emily Rudge, Rosie Storey and the
Aster team.

Illustration and design:
Andreas Brooks, Joel Galvin and Maximillian Perchik.